Scientists in Search of Their Conscience

Springer-Verlag Berlin · Heidelberg · New York

Scientists in Search of Their Conscience

Raymond Aron · Hendrik B. G. Casimir · Friedrich Cramer
Michael Feldman · Wolfgang Gentner · Michael J. Higatsberger
Aharon Katzir-Katchalsky · John C. Kendrew
Théo Lefèvre · Ole Maaløe · Chaim L. Pekeris
Albert B. Sabin · Jean-Jacques Salomon · David Samuel
Altiero Spinelli · Léon Van Hove · Siegmund G. Warburg
Victor F. Weisskopf

Edited by
Anthony R. Michaelis and Hugh Harvey

Springer-Verlag Berlin · Heidelberg · New York 1973

The help of Mrs. Renate v. Bamberg, Mrs. Irene Gillespie and Mrs. Ethel Lyons in the preparation of the manuscript is gratefully acknowledged by the editors

ISBN 3-540-06026-X Springer-Verlag Berlin · Heidelberg · New York
ISBN 0-387-06026-X Springer-Verlag New York · Heidelberg · Berlin

This work is subject to copyright. All rights are reserved, whether the whole or part of the material is concerned, specifically those of translation, reprinting, re-use of illustrations, broadcasting, reproduction by photocopying machine or similar means, and storage in data banks.
Under § 54 of the German Copyright Law where copies are made for other than private use, a fee is payable to the publisher, the amount of the fee to be determined by agreement with the publisher.

© by The European Committee of The Weizmann Institute of Science Switzerland, Zurich 1973, Hügelstrasse 6

Printed in Germany. Library of Congress Catalog Card Number 72-90441

Typesetting, printing and bookbinding: Universitätsdruckerei H. Stürtz AG, 8700 Würzburg

Proceedings of a Symposium on
The Impact of Science on Society
organised by The European Committee of
The Weizmann Institute of Science

Brussels, June 28—29, 1971

Sponsoring Committee:

Théo Lefèvre, Ministre à la Politique et à la Programmation Scientifiques, Brussels, Belgium

Altiero Spinelli, Member, Commission of the European Communities, Brussels, Belgium

Prof. H. B. G. Casimir, Leyden, Holland
Prof. Manfred Eigen, Göttingen, Germany
Prof. John C. Kendrew, Cambridge, England
Prof. André M. Lwoff, Paris, France
Prof. Tadeus Reichstein, Basle, Switzerland
Prof. Hugo Theorell, Stockholm, Sweden

Dedicated to the Memory of
Aharon Katzir-Katchalsky

Preface

"Scientists in Search of Their Conscience" is the edited proceedings of the European Symposium on the effects of science on society held in Brussels in 1971. Organised by The European Committee of The Weizmann Institute, Israel, the Conference gave scientists from varied disciplines and many countries the platform from which to explore in depth the dilemma facing them. The dilemma is the responsibility of scientists for society's use of scientific findings. Though no hard and fast conclusions were reached—in fact quite the contrary—the discussions left no doubt that scientists were becoming aware that they can no longer claim that the pursuit of knowledge is divorced from its use. Yet should they begin to face the responsibility for the application of their work, it is clear that their freedom will be impaired. The loss of freedom is of course part of the dilemma of science.

Contents

Official Opening, Monday, June 28, 1971 1
Chairman: Albert B. Sabin 3
Speakers: Théo Lefèvre 7
 Altiero Spinelli 11

Morning Session, Monday, June 28, 1971 13
Chairman: John C. Kendrew 15
Speakers: Friedrich Cramer "Can our Society meet the Challenge of a Technological Future?" 19
 Aharon Katzir-Katchalsky "A Scientist's Approach to Human Values" 33
Discussion 46

Afternoon Session, Monday, June 28, 1971 61
Chairman: Hendrik B. G. Casimir 61
Speakers: Léon Van Hove "Physical Science in Relation to Human Thought and Action" 63
 Chaim L. Pekeris "The Impact of Physical Sciences on Society" 73
Discussion 86

Evening, Monday, June 28, 1971 105
Chairman: Siegmund G. Warburg 107
Speaker: Raymond Aron "Evening Address" 111

Morning Session, Tuesday, June 29, 1971 129
Chairman: Wolfgang Gentner 131
Speakers: Ole Maaløe "Can Ideas from Molecular Biology be applied to Economic and Social Systems?". 133
David Samuel "Science and the Control of Man's Mind" 143
Discussion 152

Afternoon Session, Tuesday, June 29, 1971 165
Chairman: Michael J. Higatsberger 165
Speakers: Jean-Jacques Salomon "Science and Scientists' Responsibilities in Today's Society" 167
Michael Feldman "Science and the Crisis of Democracy" 183
Discussion 190

Conclusion 193

Speakers: Victor F. Weisskopf 193
           Albert B. Sabin 203

Appendix The Weizmann Institute 213

Name Index 217

Subject Index 220

Official Opening

Monday, June 28, 1971

Chairman:

Prof. Albert B. Sabin, President,
The Weizmann Institute of Science, Rehovot

Professor Albert B. Sabin became the president of the Weizmann Institute of Science in 1970 after a career of 40 years in medical research, the high point of which was his discovery of the oral polio vaccine that bears his name.

He was born near Bialystok (Russia), now Poland, on August 26, 1906, and emigrated to America in 1921 where he studied at New York University, receiving his M.D. from there in 1931. For thirty years he was on the staff of the Children's Hospital Research Foundation at the University of Cincinnati, during the last nine years of which he served as Distinguished Service Research Professor at the University. Since 1962 he has devoted his entire time to studies of the possible role of viruses in human cancer.

Chairman: Prof. Albert B. Sabin, Rehovot

We of the Weizmann Institute and its European Committee, who have arranged this meeting, are very honoured that so many distinguished representatives of science, world affairs and other activities have come here to explore together the impact of science on society. I did not have much to do with organizing this symposium. Dr. Josef Cohn of the European Committee deserves the credit for the tremendous amount of work involved. To me, the reason for this symposium is to symbolize the growing association between the Weizmann Institute of Science in Israel and various European organizations, without in any way diminishing the older and ongoing association with the United States of America and Britain.

If one wants to strengthen the bonds between the scientists of the Weizmann Institute in Israel and those in the international communities of science, especially in Europe, why, one may ask, choose as a subject "The Impact of Science on Society"? I must say that when I finally saw the title in its present form, I said to myself "What, another one of those debates? How many have I listened to already? Can anything new be said on the subject of the impact of science on society"?

Then, as I thought about it, it seemed that the ongoing discussion is perhaps the same as a developing musical expression. It is as if after Vivaldi someone said, how could anybody do anything more—and then Bach came along. Or, in another vein, one might be drawn to conclude that since the philosophers of more than two thousand years ago searched for the meaning of good and evil, for the meaning of life, and for the question of values—what is the point of going on?

We know this is not the first symposium on this theme, nor will it be the last. Perhaps future symposia may take another form because of growing pressures from society. They may concentrate more on the impact of society on science, for this is being felt in ever greater measure. This symposium will, I hope, prove to be a sincere attempt to find new directions for scientific exploration. At the same time it will be an opportunity to re-examine the status of scientific endeavour in relation to the society that supports and is increasingly questioning the relevance of science in the important problems facing mankind.

After granting that science is continuing to broaden our understanding of the miraculous universe of which we are a part, society none the less is asking science two questions with ever greater frequency.

The first is: "What are you doing *to* us with all the knowledge you have discovered?"

The second is: "What are you doing *for* us to help us solve the serious problems facing mankind?"

It reminds me of the inaugural speech of the late President Kennedy. In calling on the citizens of the United States for a special effort, he said: "Ask not your country what your country can do for you but ask rather what you can do for your country."

This, I think, is the challenge that society is putting to science now, saying "Don't ask us what we can do for you; but ask rather what science can do to help society."

When society challenges science in this way, my answer as a scientist is: We must work together. It is no use saying we have got to do something for you without your telling us what you think is most important and what values you hold most important. Science cannot organize adequately a major part of its effort to help society solve pressing human problems unless society defines more precisely the questions for which science is equipped to find solutions.

Official Opening

It is my hope that out of this symposium will come some of the answers we are seeking. At the same time I have no doubt that more questions will be formulated than answers will be provided, for that is the method of science.

It is now my great pleasure to call on Monsieur Lefèvre, the Minister of Science Policy and Programming in Belgium to whom we are most grateful for his help in organizing this symposium.

His Excellency Théo Lefèvre, then Minister of State in Charge of Politics and Scientific Programme in the Belgian Government, was born on the January 17, 1914 in Ghent. He was educated at St. Joseph and St. Liévin Colleges in Ghent and received his doctorate of law from the University of Ghent in 1937. In 1940 he became a lawyer to the Court of Appeal in Ghent. He has been a member of the Chamber of Representatives since 1946 and in 1971 became a member of the Senate.

H. E. Théo Lefèvre, then Ministre à la Politique et à la Programmation Scientifiques, Brussels

It is with friendship, joy and pride that I speak to you at the opening of your symposium, to welcome you to Brussels. Not so long ago an Israeli scientific mission came to Belgium after a Belgian mission had visited Israel. We are delighted at these contacts and the bonds of friendship forged on these occasions. Your symposium will, I hope, serve equally the friendship and the thirst for truth which are the two bases of all true humanism.

Eminent men will take the floor after me to touch on different aspects of a subject I hold dear, *The Impact of Science on Society*. It is quite true that scientific development is the most striking phenomenon of this century. About a million men devote their lives to it today as compared with a mere handful in other days. Their discoveries upset the economy, transform international relations, change cultures, give a new meaning to war and peace, and drive circumscribed man ceaselessly forward to ask questions of himself.

Governments and public opinion have not always evaluated this impact. The desire for economic growth, the demands of national defence, the taste for prestige have often led them a long way from all economic considerations, from all discretion which the respect for certain forms of social life and certain cultures should impose. Is it for our good or evil? Who can tell us? But what I do know is that men and peoples, drawn upwards by that spiral of technical progress, have adopted habits and ways of life that they would not wish to exchange for what they had before. An irreversible change has been brought about. In its march many things have been laid waste.

Today some among us are trying to draw up a balance sheet; hundreds of species of animals driven into extinction, immense arable and forested spaces ravaged, rivers and coastal waters polluted, human groups wiped out as if at some great turning point humanity had at one fell swoop destroyed a whole series of alternatives, of possibilities provided by nature and by history.

And at the same time, elsewhere, humanity finds itself bound to discover solutions to the problems it has created by the course it has followed. Thousands of lives saved by medicine and hygiene make thousands of mouths to feed, a first consequence of our post-industrial society. But there are others. The unchangeable movement towards the concentration of population in towns bursts our already full urban centres, the extreme rapidity of visual communication transmits the gesture before its real message and turns the problem of the education of the masses into a burning one.

Moreover, the world is weaving itself into a web of artificial objects which provide their own myths, as did in other days the springs and dark woods for the Celts – myths into which modern man throws himself without reflecting. In this day and age we need men who do reflect and think. And as in the days of the Renaissance, which saw at one and the same time the rebirth of the arts and technical skills, we need the forging of a new humanism. I would not, like some people, go and seek this rebirth in some pastoral idyll, rejecting technical expertise. Nor do I think it will be found in the aesthetics of the stretched and distended forms of electronic installations and masses of concrete.

The consequences I have raised are the realities of our time. Without new progress in science we shall not be able to surmount them. We must therefore intensify research activity still more. Now such activity can only increase if economic development continues. We must keep economic growth in its rightful place among essential objectives, but it is by no means the only one. It is in itself to be taken as only a condition of social progress. The important thing is to know what kind of society we wish to shape. To the question of the impact of science on society, we should one day be in a position to respond by the orientation of science for society. But what kind of society?

During the last ten years, scientific policy has been preoccupied with increasing the budgets devoted to research and to improving their effectiveness. For us as Europeans, the rift between the United States and our countries has been the *leitmotiv* of our thoughts and the basis for our actions. It remains so, but it is no longer the only one. Scientific policy tries more and more to pinpoint the aims and objectives of success in technological innovation.

If the aims and objectives must be defined it is because they can no longer be taken for granted. The evidence is no longer assessed for us. That is why we must ceaselessly question the effect of human actions on the natural and artificial environment, interactions which always risk compromising the social and human equilibrium that is our ultimate objective.

I can already foretell that this symposium will permit you, Mr. President, Your Excellency, Ladies and Gentlemen, a most fertile exchange of views, and I hope that each one of you will return from it even a little more enquiring than when you came. I thank you.

M. *Altiero Spinelli* is a Member of the Commission of European Communities in Brussels. He has been active in the movement for a united Europe for many years having founded the Movement for a Federal Europe in Milan in 1943. He was born in Rome on August 31, 1907 and educated at the University of Rome. From 1962 to 1966 he was Visiting Professor at the Bologna Centre of the School of Advanced Studies of Johns Hopkins University. In 1966 his book *The Eurocrats* was published by Johns Hopkins University Press and he has many other publications on this subject including *Problemi della Federazione Europea* published clandestinely in Rome in 1944.

M. Altiero Spinelli, Brussels

I bring you greetings, brief but cordial, on behalf of the Commission of the European Community. The fact that this symposium takes place in a city which is not only the capital of the Kingdom of Belgium, but also the capital of a Europe in formation, is a homage which science pays to Europe. For science today is without doubt universal, it knows no frontiers, and also it had its cradle in Europe.

The theme of your symposium, *The Impact of Science on Society,* is an important and serious one, for as Monsieur Lefèvre reminded us just now, once science had above all a predominantly cultural aspect. Today it has also an economic value, a social value, a political value and a moral value of the first order. For you it is not necessary to dwell on each one of these points; it suffices to remind ourselves that all this necessitates a new and profound examination of what science can contribute for good or ill, and of what it can gain of good or ill from society.

Allow me simply to linger on one point. Science has become more and more expensive and sophisticated, research more complex, so that from now on it is no longer an affair of individuals or small groups, but has become a political matter of the first order in which all governments are interested.

Expenditure for scientific development can no longer be regarded as something which develops in itself independently of other needs. There is a necessity to regroup the initiatives of researchers, of governments, that is to say it is more and more necessary to have planning of scientific research from the simplest level within any single research centre, right up to world level. Planning means the establishment of priorities between the needs of research and other needs; it means a continuous struggle to evaluate the interests of science and to demonstrate that they are useful to the development of society.

In this struggle it appears more and more that we must concentrate on a problem that is becoming ever more serious for all peoples, the problem of research into improving the quality of life, into the quality of our society rather than into the quantity of needs to be satisfied. This subject is also a matter of interest and reflection for the European Community.

The Community has until now been present in the realm of scientific research in a slightly accidental way, thanks to the chance formation of its institutional structures. It has therefore concentrated on only a few problems of scientific research. But as the Community develops, and it is well on the road to development, as the Community tends to unify in a common destiny the free peoples of Europe, the Community must necessarily place at the centre of its preoccupations, to a greater and greater extent, the problems I have just cited and which are now under your scrutiny.

All this explains the interest which the Community takes in your work and the wish that I formulate for the success of this symposium.

Morning Session

Monday, June 28, 1971

Chairman:

Prof. John C. Kendrew, Cambridge

Professor John C. Kendrew is Deputy Chairman of the Medical Research Council Laboratory of Molecular Biology, which started as part of the Cavendish Laboratory at Cambridge, and he is Director of its Division of Structural Studies.

He was born on March 24, 1917 in Oxford and entered Trinity College, Cambridge, as a Major Scholar in 1936, graduating in Chemistry in 1939. During the war he was on the staff of Sir Robert Watson-Watt and worked on radar. After the war he became interested in biology, returning to Cambridge in 1946 to collaborate with Max Perutz on the structure of proteins and in particular myoglobin. For this work he was awarded, with Perutz, the Nobel Prize in Chemistry in 1962.

Chairman: Prof. John C. Kendrew, Cambridge

I have very great pleasure in acting as Chairman of this Opening Session of our conference, especially because of my own long association with the Weizmann Institute, one of the most distinguished centres of pure scientific research in the world. In addition, the Institute has, and I think always has had, a particular concern with problems of the relationship between science and society. So it is, in my opinion, a very appropriate thing that the Weizmann Institute should be organizing a meeting on these topics.

There are many reasons why I think it would be impossible for a scientist in Israel not to be concerned with society at large and his own relationship to it. For all these reasons, the problems of society press in on every side and cannot be neglected even by the most ivory tower researcher. Perhaps the only difference between Israel and Western Europe in this respect is that in Israel the problems are more obvious; in fact, for us in Western Europe the problems, even though they may not always seem so obvious, are just as pressing and in the last few years they have become all the more urgent.

If we look back to the happy period in the 1960's when governments and scientists seemed to assume "the more science the better", all you had to do to get more money was to ask for it. On the whole governments were rather generous. It is only in the last three or four years that the questions from governments have become louder: why and for what purpose do we supply this money? Is the work we ask the scientists to do relevant to the problems of the people who pay for it?

The fact is that scientists are not used to having to justify their own activities. To justify a grant application for a particular research project

—yes; but to justify the whole scale of scientific activity — no. They assume that society is ready to pay them to do what they enjoy doing, and that economic benefit automatically follows from it. These assumptions have come more and more to be questioned by governments and by society at large. A number of studies of the connexions between academic, "pure", research and the large-scale technologies that come later have shown that the links are often not directly causal and are sometimes quite tenuous; certainly cost-benefit analysis cannot, or at least up to now has not, provided a justification for the actual amount of money a country spends on research.

If science were to be thought of simply as a cultural activity, like music or the visual arts, it is evident that scientists would be much less well-off than they are today—just compare science budgets in any country with budgets for the arts. If scientists ought to be given more resources, how much more should they get? What they happen to get today? Twice as much? Half as much? No-one has yet succeeded in treating these matters quantitatively, yet the questionings become louder and in many countries there have been serious cut-backs of resources.

Science has also been criticised from a quite different point of view than the economic one. To what extent does the advance of science bring with it dangers as well as benefits?—dangers to human life and to the environment. There have been questionings, of course, from many young people, some of whom go so far as to say that in their opinion scientific research is a dangerous activity and should be stopped. And the questionings have come from ordinary people too. I think that perhaps in biology these questions seem even more urgent now than in some other fields, such as in physics and chemistry where it has been recognized for a long time that there are dangers, as well as benefits, from the applications of pure science.

In biology people have only recently become aware of the fact that although pure biological research appears on the face of it to be conferring absolute advantages, by, for example, reducing infant mortality, this apparent good brings with it problems of a population explosion. One could go on giving examples of this kind for a long time. It is perhaps therefore fitting that we open our meeting with two distinguished speakers who will talk to us about the way they see the relationship between their fields and social problems in biology.

I want to introduce our first speaker, Prof. Cramer, Director of the Max-Planck-Institute for Experimental Medicine at Göttingen. He is a chemist, he is a molecular biologist of a very pure kind perhaps, but he is also a director of an institute for experimental medicine. His interests spread widely in different areas of science and as the name of his institute indicates he has a concern with the application of biology. So I think that it is appropriate he has been invited to talk to us. His subject is: "Can our Society meet the Challenge of a Technological Future – the View of a Biologist."

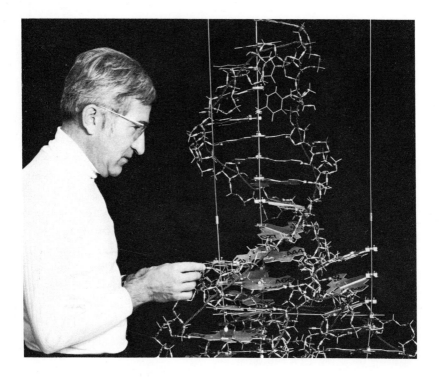

Professor Friedrich Cramer is Director of the Chemistry Department of the Max-Planck-Institut for Experimental Medicine, Göttingen. He was born in Breslau on September 20, 1923, and obtained his Ph. D. from the University of Heidelberg in 1949. During the years 1953–54 Professor Cramer was guest lecturer in Organic Chemistry at Cambridge University and from 1959 through 1963 Professor of Chemistry at Darmstadt University.

He is known for his work on enzyme mechanisms, on nucleic acid chemistry and nucleic acid structure.

Can our Society meet the Challenge of a Technological Future?

Prof. Friedrich Cramer, Göttingen

Since man emerged from the twilight of the pre-hominids, the first and only being with critical awareness, he has been attempting to order the world with his awareness and understanding, and to attain power over nature through his knowledge. At the beginning this process had to content itself with a largely symbolic check on the environment in accordance with the small surveyable region and with the forces to which man was exposed: his nature divinities and demons are psychic projections of this process.

Out of the purely symbolic method of observation, somewhat like that found in Ancient Egypt, comes the description of facts and the experiences of human groupings, public affairs and interactions. Beginning with Greek historians, the occidental view of history arose from the eschatological orientation of Near East religions. These assigned an inherent transcendent destiny to history, first perceived as the history of God, as the progression from Creation to Salvation. An attempt to undertake a secularized historical ontology was made in the 19th Century by Hegel, and by Marx based on Hegel.

Eventually the refined methods of scientific observation and the consistent application of the Principle of Causality, plus the possibilities of comparative biology, led to the realization that all living species can be ordered into a family-tree system, and that they developed from primitive to higher forms. The unfolding of forms of life up to *Homo sapiens,* involving a time lapse of barely one billion years, is an established scientific fact today.

Nevertheless, one matter remained inexplicable and had to be ceded to the Creation story, that is, the development of a living single cell (microbe). From that point on, Darwin's system of survival of the fittest (Natural Selection) functions flawlessly in all branches of biology. Now biochemical and molecular-biological research are involved in inserting the last remaining link of this evolutionary series from anorganic crystals to *Homo sapiens*.

We can form a detailed idea of the origin of life on our planet based on our knowledge of global history, and on experiments that can be repeated. In recent years we have reproduced in a flask the earth's atmosphere as it was approximately one billion years ago, and the conditions presumed to have been present then. When a scientist puts purely anorganic substances into such a flask, there appear in time the vital building blocks of which currently living organisms are composed: the nucleic acid components and amino acids, the most important ingredients in proteins.

On the basis of these experiments it can be assumed that these compounds were formed on the earth's surface at a point in time when the ammonia-CO_2 atmosphere began – in one place – to shift to our present oxygen-nitrogen atmosphere. Substances could have been formed in this primordial liquid that catalysed their own reproduction. Deoxyribonucleic acids (DNA) are the carriers of inherited characteristics in all living cells. These nucleic acids are a long, complementary double strand called the "Double Helix", which contains the entire genetic information in a four-letter transcription enciphered according to the genetic code. The deciphering of this code represents one of the century's greatest accomplishments.

The double strand of nucleic acid contains the genetic information in duplicate. It is copied completely and correctly at the moment of cell division. This occurs flawlessly in our time with the help of certain cellular mechanisms. Can a nucleic acid which accidentally developed in a primary substance have a selective advantage over another chance-occurring nucleic acid, so that it is ultimately selected out? This is quite conceivable: Natural Selection need not ensue with regard to criteria which seem important to us today. In the prebiotic phase, certain sequences might have had an advantageous origin because by feedback they catalysed their own formation. Even if this formational advantage consisted only of a fraction of one per cent, the total nucleic acid population eventually could be standardized to the one nucleic acid, since all other nucleic acids would arise only statistically.

A reciprocal (or mutual) action must have occurred between amino acids (or primitive proteins) and nucleic acids in the prebiotic phase so that the nucleic acids could somehow organize the construction of certain protein materials. We know exactly how the nucleic acids, acting as organizational elements of the nucleus, regulate the entire protein synthesis and how the linear information of nucleic acids is determined and translated in a definite protein sequence. This process comprises a most sophisticated mutual action between the nucleic acids and proteins, in which the so-called transfer ribonucleic acids (RNA) play a decisive role. This molecule, by the way, happens to be the subject of my research.

The microbe (monocellular creature) is found at the end of the prebiotic phase. The molecular processes up to this stage of development have been essentially understood. Recently they were dealt with quantitatively in a novel physio-mathematical theory by Professor Manfred Eigen. Life is characterized by two conceptions:

1. Reproductory autonomy, i.e. the capacity for autonomous self-preservation in the environment, and of reproduction.
2. Evolutionary teleonomy, or "ultimate aim", i.e. the (inner) laws according to which the biosphere is changed, in an ordered manner, towards an end.

We can comprehend to a large degree the autonomy with its molecular reproducing mechanisms. But what is meant by "Teleonomy of Bios"? It is *the meaning of biological history,* the history of evolution. Biology thus is connected to universal history.

The question of its teleonomy or ultimate aim has to be posed according to the same principles and answered just as the question about the meaning of universal history. Accordingly, biology is subject to all speculations which govern the historical theories and ontologies. These theories can be divided into two groups of contradictory solutions:

Teleonomy enables and regulates autonomous reproduction;
or
Teleonomy is generated by self-reproductivity as a "super-structure" *(Überbau).*

How can this dilemma be solved? Classes with the former group of solutions are all vitalistic theories proceeding from the idea that every substance has a *vis vitalis* at its disposal. One metaphysical solution is the assumption that reproductivity and "Directedness" are inherent life qualities that express themselves more or less vigorously depending on their evolutionary rank, and in which man has reached the highest rank. Man, or at least a future human that has achieved maximal transcendent reason, is the final product of evolution. This is the interpretation of Teilhard de Chardin.

Such theories, whatever their inclination, are not scientifically relevant because they employ axioms from non-scientific fields. The latter solution, namely that teleonomy is the superstructure of reproductivity, is the only physically (scientifically) defendable viewpoint. The thermodynamics of irreversible processes do not rule out the notion that, contrary to statistical probability, at particular points in the universe highly complex structures have formed and organized themselves autonomously on an increasingly high level, creating their own history. The significance of this history need no longer be questioned: it is a constant, inevitable *"Flucht nach vorne"* (escape forward) which generates organisms, brains and thoughts.

Hence evolution no more demands a *causa vitae* than Marxism required a *Zeitgeist* (a spirit of the age). Such a *causa vitae* is not included in the prerequisites of the scientific system; rather it is excluded as a pre-condition. Consequently, it cannot reappear at the conclusion.

Let us consider a consequence of the thesis that the reproductivity of Bios can generate its teleonomy. An unequivocal directional indication of evolution towards a definable goal, in the sense of a superstructure arising from the reproductivity, leads to a consistent and unambiguous result only if the reproductivity itself is unequivocal or, in other words, if biological evolution can proceed in one direction only, i.e. if there is only one evolution or only one evolution is conceivable.

Nothing justifies the assumption, however, that there cannot be several such mechanisms. Indeed, it would correspond flatly to the non-permissible introduction of a *vis vitalis* if such an evasion were required.

Parallel or diverging evolutions are conceivable in any given self-contained systems which can generate their own teleonomy and their own natural history. The possibility of diverging evolutions, e.g. in various planetary systems, can therefore not be excluded. Admittedly, it cannot be proved, but the former fact alone suffices to rule out one autonomous evolution with a generally valid "ultimate aim" or teleonomy issuing from it, because the latter is most improbable. It is not acceptable according to scientific method to elevate a very unlikely hypothesis to a theory.

For example, instead of a system based on individual brain power, insect nations with collective abilities might have developed. Consequently, there is no generally valid natural history. If we cling to the notion of natural history at all, we must verify the inherent relativity of natural history. What is the story with our particular evolution, namely the generating of *Homo sapiens*?

In the Early Stone Age, during the transition from nomadic hunters to sedentary farmers, man departed from evolution in the Darwinian sense. Since then he has no longer been subject to the law of survival of the fittest. He learned to avoid hunger by storing food; cold by building fires; self-destruction by introducing definite moral laws; and germs through hygiene. Man's total hereditary factors were no longer a result of the struggle for survival. The hereditary factors have been stagnant since Gilgamesh, Theseus, or the *Nibelungen* even with regard to the instinctual pattern. By handing down acquired knowledge and by means of communication, man ruled nature in a manner that not only brought his evolution to a close, but also a great deal of the evolution in the natural environment.

In the latter half of the Twentieth Century these changes and revolutions will be effected mainly by biological research. The vision of the final phase of organ transplants is already known. According to this view, by degrees all deteriorating organs up to the longlived, potentially immortal brain would be removed from aging people and replaced by organs from anthropoid apes. The personality, the seat of which is in the brain, with its faculties for thought, association and memory seated in the brain would be preserved.

As an example, an individual renewed in this way could fill a particularly important office and do so especially well because he would possess experience transcending generations — though of course he would not sell well as a political candidate on TV. In a further step of this vision of the future, natural organs could be dispensed with entirely, the brain being furnished with nourishment, oxygen, hormones and all the necessities with the help of the veins and their access means.

This would give rise to a new kind of ancestral gallery, in which people would be able to locate their ancestors on "care poles", chat with them about times gone by, and project for them the "Newsreel" of the present. This assumes that there will still be ancestors, for by that time natural propagation and parenthood presumably could have been replaced by more objective and impersonal methods.

Such visions — which may even contain the seeds of realization — are apt to conceal the power and significance of biological research because they steer one's attention toward distant possibilities, and away from its present achievements. In the Nineteenth Century, when the first nutritional catastrophy of modern times threatened to break out because of population growth, Justus Liebig invented inorganic fertilizers based on biochemical research. At that time, undernourishment was an actual danger for Western Europe, out of which, for example, arose the ideas of Malthus.

The breathtaking advances of medicine are the result of the biological interpretation of medicine. For example, the discovery of bacteria by microbiologists rendered possible the combating of infectious diseases. From biochemists' elucidations of the metabolic processes and their pathological alterations followed therapy, e.g. for diabetes and pernicious anaemia. The study of the rudiments of nutrition brought about the attenuation of infant mortality. The explanation of sexual hormones by chemists and biochemists ultimately led to hormonal contraception; the discovery of blood groups brought about blood transfusions; and the clarification of immune reactions will someday ease organ transplanting. Even though biology has not had its "atomic explosion", we nevertheless find ourselves in the midst of a population explosion, resulting from biological research.

I cannot predict which problems will be solved when. However, it is certain that some of these questions, once they are resolved, will work themselves deeply into human society. In a feedback process, man fashions the world through research in a manner that creates new obligations when the old ones are eliminated. That introduces irreversible developments whose consequences have not been initially comprehended and are by no means desirable. Suddenly man finds himself governed by conditions to which he indeed gave rise but never wanted.

Can scientific progress continue in its present manner? The dynamics of progress in research, economics, prosperity, and in all facets of life were studied at the beginning of the century by Henry Carter Adams, who advanced the Law of Acceleration. Adams ascertained that in the 19th Century the generation of energy (at that time essentially from coal output) doubled every ten years. From 1500 to 1799 this doubling took place every 25 to 50 years. Using this outcome, one can extrapolate for the future. Acceleration phenomena of this kind arise through positive feedback in all branches of natural science.

Progress is bound to definite material pre-suppositions. The positive feedback of progress has simultaneously an element of self-containment. Whenever there is a limit to progress somewhere—and there must be a limit—exponential acceleration signifies that this limit is being approached with increasing speed. This is evident for some typical parameters: the world population cannot increase unchecked; energy consumption cannot keep multiplying if only because the by-product heat will become intolerable; the *per caput* income must eventually reach a ceiling, because not everyone can own everything.

It is clear beyond doubt that the accelerating kinetics of progress exclude the possibility that progress will be an eternal attribute to human history. For example, the increasing number of those participating in scientific research is so far above the acceleration rate of population growth that by the year 2050 everybody will have to be a scientist! This conclusion is naturally incorrect; however, it signifies that the progress of research must slacken.

I would also like to consider the question from a psychological viewpoint. The psyche of modern man finds itself in a process of metamorphosis. Attitudes towards erotic and instinctive behaviour have changed, in part at least, as a result of biological research. Since Freud we have known what effect the structure of Eros has on society; Herbert Marcuse summarized it in *Eros and Civilisation*.

The Performance Principle which has yielded progress is a subject questioned today. When it is abolished, progress will stop. It is possible that mankind will have reached such a standard of life in a few generations that it will be able to afford a complete progress-freeze. Progress is, notwithstanding, the *competitive principle of Faustian Man*.

The end of progress is only possible if all people, human groupings and nations, including the Chinese, resolve that they have "had enough". Even if *one* tries to break out, the metastable equilibrium is disturbed. Only through a new *contrat social* of a higher order can progress be ended. Can one envisage such a global "Non-proliferation pact", ensuring non-proliferation not only of atomic weapons but also of economics, population and pollution? Is the furious acceleration leading us to a Utopian Golden Age or into the general chaos of the struggle for survival on a higher order between social groupings, nations or continents?

We are faced by a dilemma. Biological research, which set out to free man from obligations, has actually eliminated the inevitability of evolution. Scarcity has not diminished, however; it has shifted and restructured itself. In the quest for the Golden Age we are confronted by the danger of complete chaos. The remedy seems to me: we enter the age of planning, for future planning is a way to bear the responsibility for progress in an integrated society.

The future is already with us, and we mould it whether we want to or not. But in contrast to earlier epochs, we are the bearers of the process of production and are ourselves the product of it. "Since man has irrevocably become a citizen of a world he himself produced, since he has possessed the power to produce not just inanimate goods, but his own world, his vital conditions, and largely his own destiny, since he has encountered himself as a product—his future has taken on another character. The future of the technological world differs qualitatively from everything that earlier epochs called future." (Georg Picht.) If this world is to be worthy of human beings, who will govern it with freedom and reason? We must design and plan the production of the coming world in a manner that allows man *humaner Raum,* humane "elbow room".

This planning has a technical aspect. A technically perfect inventory must be carried out. All facts necessary for planning must be compiled, worked through scientifically and alternative suggestions for individual partial questions submitted. An abundance of planning suggestions arises according to the tremendous existing material possibilities for every situation. We are living not only in an affluent society, but also in a society with a surplus of potentialities.

After material progress in the affluent society begins to overcome scarcity and its associated asceticism, it becomes evident that a new asceticism of human potential is imperative. This is a task which transcends technology, a task which pre-supposes a humane model on to which man projects himself. Never before has mankind faced such a situation: the absence of want and repression. And at the same time, a scarcity, a repression of a higher order appears on the horizon.

Thus one can envisage overcoming this situation, as Herbert Marcuse considers it possible for the liberation of Eros at the summit of civilization: "It would still be a reversal of the process of civilization, a subversion of culture—after culture had done its work and created the mankind and the world that could be free... Under these conditions, the possibility of a non-repressive civilization is predicated not upon the arrest, but upon the liberation of progress—so that man would order his life in accordance with his fully developed knowledge, so that he would ask again what is good and what is evil. If the guilt accumulated in the civilized domination of man by man can ever be redeemed by freedom, then the "original sin" must be committed again: 'We must again eat from the tree of knowledge in order to fall back into the state of innocence' (Heinrich von Kleist, *Über das Marionettentheater)"*.

Scarcity is the biological norm, plainly speaking, the motivating force of the living. The elimination of need causes grave and probably even fatal disturbances if it is not accompanied by the appropriate consciousness which we know as a form of asceticism or voluntary renunciation. The planning potentialities show us the necessity of a New Asceticism. The technological planning task directed toward a humane future is a task for which science supplies the material, but which cannot be resolved by science. Man transcends his horizons in contemplating the future.

Chairman: Prof. John C. Kendrew, Cambridge

We are very grateful to Prof. Cramer for painting this sweeping canvas starting with the primeval soup and then through the mechanisms of molecular biology to the philosophical implications of the evolutionary process. Incidentally I noticed his taking a kind of side swipe at neo-vitalism; — it is perhaps significant and a little sinister that it seems necessary to take a side sweep at vitalism, but probably it is. He cited the modern achievements of medicine, and the H.G. Wells type phantasies which could become realities in a more distant future if we allow them to. This was Prof. Cramer's view of a kind of challenge which society has to meet when faced with the potentialities of biological technology.

There is the other side of the coin as well; can our science and our technology meet the challenge posed by society? And as I understand it, this is the theme of our next speaker. His title is "A Scientist's Approach to Human Values". Prof. Aharon Katzir-Katchalsky is a senior member of the staff of the Weizmann Institute. For those few of you who have not had the pleasure of knowing him for as long a time as I have, may I say he is one of the few people at the Weizmann Institute who over many years has been involved in building it into the place it is today; he is one of the living embodiments of the Weizmann Institute.

I will not recount his many distinctions, though perhaps I may be permitted, since we are having our discussion in Brussels, to mention just one — his Honorary Doctorate from the Free University of Brussels. He is Head of the Polymer Department at the Weizmann Institute, and he is also a biologist. So he will talk from the point of view, at least in part, of a biologist on a scientist's approach to human values.

Professor Aharon Katzir-Katchalsky was at the time of this conference Head of the Department of Polymer Research at the Weizmann Institute of Science. Born in Poland in September 1914 he came to Israel in 1925 and studied at the Hebrew University of Jerusalem.

He had been a visiting professor at a number of universities in Europe and the United States, including Harvard Medical School and the University of California at Berkeley. Dr. Katzir-Katchalsky specialised in the field of the thermodynamics of irreversible processes and especially its application to biological membranes. He is the inventor of a mechano-chemical engine, the first robot with muscles.

Dr. Katzir-Katchalsky was one of the many innocent victims who lost their lives during an attack by terrorists at Lod Airport, Israel, May 30, 1972.

A Scientist's Approach to Human Values

Prof. Aharon Katzir-Katchalsky,
The Weizmann Institute of Science, Rehovot

Not only is it difficult to follow after the inspiring speech of the previous lecturer, but I must confess I had great difficulties in preparing this lecture. For a scientist it is easy to talk about his own field for he is confident he knows what he is talking about. But when he transgresses the realm of his activity and discusses general human problems, he finds himself a dilettante.

In particular, I had difficulty in deciding what an Israeli can tell an international audience. Israel is not in the midst of a striking scientific development; the social problems arising from the development of technology do not have such a clear impact on Israel as on other countries.

The justification for this lecture is that Israel is undergoing a rapid process of economic growth, and many problems of our time can be studied by taking Israel as a model. During the 23 years of the existence of the State, our scientific manpower has grown by a factor of 10, while the science-producing population has grown only by a factor of 4. The process of scientification is so rapid that following the calculation, which Prof. Cramer has mentioned of Derek de Solla Price, in another hundred years there would be more scientists in Israel than its total population!

The impact of science on Israel has brought out a number of human problems which are not solved by the fact that it helps science-based industries and raises the general standard of living. For example, about half of the population of Israel is made up of immigrants from underdeveloped countries for whom science is a foreign language. Generally, their income is quite good, and they cannot complain that economically they are in the position of the under-dog.

The very fact that the country is becoming more and more scientific creates in them a feeling of deep frustration; they recognize that they cannot catch up because science is talking a language they do not understand. Science, despite its international character, speaks a Western language. And with all our efforts, we have not found a way of translating science into a universal language so that it could be incorporated without destructive side effects into cultures which have nothing in common with those of the West.

There is another detrimental aspect of the growth of science in Israel. The founding group of the State of Israel and the backbone of its early development was the pioneering movement. This is a puritan movement, based on a rather naive Tolstoyan socialism, the value system of which carries a strong imprint of classical Judaism.

The scientification of Israel imposes on the pioneering movement, in all its ramifications, a new outlook and a new scepticism on the meaning of life. As expected, the primary effect is a deterioration that threatens the very existence of one of the vital elements of the nation. In this example we recognize that the famous neutrality of science is neutral only in its approach but not in its consequences.

The "neutrality" of science reminds me of the joke about the Rabbi who had to sit in judgment. One party approached him, pleaded at length and convinced him, so that he declared: "You are right..." Then the other party jumped up and shouted: "What do you mean he is right—I have such weighty arguments..." He brought up these weighty arguments and also convinced the Rabbi, who stated: "You too are right..." The wife of the Rabbi then tackled him: "This is nonsense! How can you say that both sides are right?" The Rabbi thought deeply, looked at his wife, and then made the profound statement: "Indeed, you too are right..." This is the type of neutrality that we expect from science, but it is rather difficult to put it into practice for ultimately one has to choose who is right. It is this avoidance of unequivocal and imperative answers that converts science into such a demoralizing power.

To be sure, the major problems transgress the boundaries of Israel, and as the previous speaker has mentioned, not only the physical but also the biological problems weigh heavily on the mind of the science-dominated society.

After the bomb, the atomic physicists suffered a guilty conscience. This did not help much, but as a result many of them turned to the "peaceful" fields of biophysics and molecular biology. Today, however, the molecular biologists are facing a situation very similar to that of the atomic physicists a generation ago: but there is still time to think and to plan. A few typical examples will make clear how molecular biology, the study of the brain and other topics in biological science, are posing heavy moral questions that require a new ethical approach.

Let us start with a relatively simple example. It is known that the human chromosomes are such that the female has two large X-chromosomes, while the male has one large X-chromosome and one smaller, weaker chromosome, the Y-chromosome. But it has been discovered that there are some abnormal cases, men who have two Y-chromosomes, i.e. X-Y-Y and, moreover, that there is a high correlation between criminality and X-Y-Y incidence. The correlation is not total, but a high percentage of all men who carry X-Y-Y chromosomes have strong criminal tendencies.

The following ethical question is now evident: should a man who has not committed a crime, but who has a X-Y-Y structure, be put in a closed institution or should he be allowed to run free until he commits a crime? The basis of all democratic societies is that a man cannot be punished until he is found guilty. In our case, the carriers of X-Y-Y chromosomes should therefore be left in freedom, but in so doing are we setting up the conditions for some horrible crime to be committed?

Or let us turn our attention to even more difficult problems: Genetics has reached a stage when one can envisage that in the foreseeable future it will be possible to control human genes and to practice "genetic engineering". It is known that Khorana has succeeded in synthesising a complete gene of a low plant, and in principle one could also dominate, change, put in or take out genes of the human species.

Assume that we shall be able to control the genes. We are then ready to tackle hereditary diseases and make the next generation congenitally healthier than its parents. But this raises the possibility of giving society the means of controlling genetics in general, of interfering with the natural processes of life and of shaping human beings at will, without providing the ruling members with a suitable ethical dictum.

Let me give you an additional example in the same field: recently the transfer of the nuclei of mature cells into an unfertilized egg has been successfully accomplished. One can take a nucleus of a mouse and put it into an unfertilized egg, the nucleus of which has been destroyed by irradiation. This transferred nucleus grows and divides within the egg, developing into an exact replica of the individual from which it was taken. This allows mass production of one type of individual.

Should the technique work in man and some crazy dictator have clever military ideas, he could make an army of replicas of the same "useful" type. This is not science fiction, and although the realization may be very far, we are not talking about fictitious ideas. One ethical problem is whether we should disclose every one of these achievements to society, or should the scientists keep their results secret?

Still another example is control of the brain. By inserting electrodes in the hypothalmic region of the brain one can stimulate certain cells, and create a state of extreme happiness. Quoting from one of the recent issues of *Time Magazine,* the subject experiences "super-sexual satisfaction".

It is also possible to make people happy by other technical devices. Thus, some time ago, a research worker in the States found that by conditioning the rhythm of the brain without inserting electrodes it is possible to achieve a state of euphoria. If these results become a widespread technique, we can readily envisage some of the consequences of putting men in a fool's paradise in which they will accept without criticism whatever is done to them.

Another group of challenging problems, which the impact of science has brought to our attention, is the following: All of us are the product of a certain attitude towards our environment. We belong to a family, we are citizens of a town and members of a nation, but generally speaking our belonging to "mankind", or our international relation, is rather weak. Present-day development of science imposes the dictum that international relations are a "must" for the survival of mankind.

Let me explain. For many centuries, maybe millenia, man regarded the globe as an open system, i.e. whatever his activity was, its consequences were absorbed by an "infinite reservoir of the surroundings", which did not mind what he did. The recent development of technology, based on science, has converted the world into a closed system; whatever is done in one country influences the surroundings and is felt also in other countries.

Pollution is not only a technical problem but is a new attitude toward the globe, an attitude toward a *finite, closed* system. Man is compelled to find his way to a new morality not only on the individual level—as dictated by molecular genetics and brain research—but on a national level, as required by the challenge of the finiteness of the global system.

Many years ago Jean Piaget studied the question of our territorial perception and made the following nice observation: He asked eight-year old Genevoise children what is the relation between Geneva and Switzerland. They generally drew two circles—one circle for Geneva and another one for Switzerland and made it clear that if they belong to Switzerland they do not belong to Geneva, and vice versa. At the age of 10 or so, the children started to draw bigger circles for Switzerland, and a small circle within it for Geneva. This signified a transition from the territorial belonging to the city to an internalization of the belonging to the country which embraced their local adherence.

The problem facing us today is to widen the circle to comprise not only our countries, but our global citizenship as well. This new morality does not contradict national belonging as it does not compete with our belonging to a city or a family; but it means that there is a hierarchy of belonging which is *internal,* an integral part of ourselves. If this stage of psychological development could be reached, then we would be fit to cope with the problems of the globe or with those of international relations.

Because our generation did not succeed in meeting the challenge of the impact of science on society, the younger generation cannot forgive us. Some time ago a book, *The Making of a Counter-Culture,* appeared which summarizes the attitudes of young people toward science. The author, Theodor Ruszak, accuses all science of being an anti-human activity, for through its impact we become strangers to ourselves and cannot tolerate our existence.

The great sociologist Durkheim said many years ago that for thousands of years man tried to master nature in order to make for himself a better life; but when at long last he mastered nature, he lost interest in life itself. This is what is happening to the young generation: their feeling is that first of all one should get rid of that natural curiosity which underlies science.

There is little doubt that the curiosity of the scientist has deep biological roots, for curiosity is found in all the animal world, in insects and in the lower mammals as well as in high primates. It is curiosity, and not some materialistic factor, which led to the development of science.

The young generation, however, accuses science of leading to a hypertrophy of curiosity. The scientist, they claim, has a pathological curiosity that satisfies only himself, and for this satisfaction he is ready to sacrifice the life of mankind. Hence they see the salvation of mankind in anti-intellectualism.

This approach is not only dangerous, but it is self-destructive, since ultimately there is no morality and no ethical philosophy without cognitive foundations. This was recognized in the book of Genesis, where there is a wonderful statement that only when man had eaten from the tree of knowledge, when he developed his cognitive processes, "did he become like one of us to know good and evil".

In other words, to select between good and evil one must have cognition, for without intellect we cannot distinguish between the two. Anti-intellectualism is therefore *a priori* anti-moral and our problem is how to foster an intellectual approach which will carry moral weight. It is the problem which I have often discussed with many Israeli youngsters, and it is surely a cardinal international problem.

For many years, scientists and philosophers looked to science as a guide toward a new way of human existence. The first were the physicists who recognized that though the laws of physics cannot direct human behaviour, the fundamentals of science should be consistent with the principles of morality. It was pointed out that science teaches man objectively, showing how to evaluate correct, honest information about the real world. Indeed, the scientific approach has a more profound meaning because it is the basis of personal maturation.

Freud recognized correctly that one of the basic principles of maturation is the acquisition of the *"Realitätsprinzip"* which underlies the differentiation of subject from object. It is perhaps significant that the hippies in their struggle against science deny the Freudian *"Realitätsprinzip"*. Thus Timothy Leary says that under the influence of LSD he is in a wonderful state because the difference between subject and object disappears. This means that LSD evokes the infantile state in which separation between the subject and the object world still does not exist. Negation of science is therefore a retreat from maturity, with all its social and moral consequences.

There is another important aspect, pointed out by the philosophers of science, namely, that science is intrinsically an anti-authoritarian system. This derives from the fact that science does not recognize the authority of man but relies for laws on the external judgment of Nature. Thus a recognized scientist of unchallenged authority may have a wonderful pet theory but if a student makes an experiment and the results contradict this theory, then the student will be right and the scientist's authority will be of no avail.

The higher authority, transcending all human authorities—that of the experiment, or, metaphorically, the authority of Nature—is a trans-personal authority which should teach scientists humility and could justify a democratic mode of life.

It is well known that scientific humility is very limited, as demonstrated by the following anecdote: A well known scientist was lying on his death-bed, and his students who surrounded him were praising his achievements, his wisdom and his knowledge. All of a sudden, they saw that the lips of the dying man were moving. They bent down to listen to his message, and this is what they heard: "While praising me, you have forgotten one important aspect—I am extremely modest too!...."

These philosophical approaches are very interesting and illuminating, but hitherto they have had no practical influence on human behaviour. Recently, the biologists came to the conclusion that since biology may enlighten us generally regarding man and the motivations for his behaviour, human biology should be consulted for ethical conclusions.

As an example, let me mention the study of "non-verbal" communication. There are many channels of communication, one of the most important being body movements, which are genetically determined. Thus Eibel-Eiberfeld found that movements of the head and neck of marriageable girls in the presence of men are fully automatic, like the ritualistic movements of birds or fishes, common to all human races and independent of cultural background. Non-verbal communication is not only an important channel for the fixation of human relations but provides an interesting clue to understanding the biological foundations of human behaviour.

Indeed, behavioural studies show that several of the ten commandments have early biological origins. Thus the taboo on incest is found already in the chimpanzees, and "thou shalt not kill" is a commandment common to all higher animals and is basic to their ritualistic behaviour. Even the elements of personal courage are observed in nature. Thus in all social insects, when danger faces the group, the individual will sacrifice himself for the benefit of society. If the movement of ants or termites is hampered by fire, thousands will throw themselves into that fire to extinguish it and let the society move on.

When analyzing the nice example of "ant courage", we readily recognize that there is a major difference between animal and man. The hero-ant has no individuality, for its behaviour is fully determined by a genetic dictum. Human morality begins at the point where the determination is incomplete and there is a choice. As long as behaviour is fully determined it cannot be regarded as ethical.

Only when there is an individual, with his personal interests, with his private tendency to survive, and only when his interests clash with those of society and he is required to make a choice to decide what is right and what is wrong, does his behaviour become ethical. This jump from fully determined to free behaviour is the transition from biologically to culturally determined behaviour and from animals to humans. As long as we do not recognize this jump we do not deal with human affairs, but stay in the realms of biology. It is the transition from biological to cultural development, from Darwinian evolution to cultural progress, which makes us human and imposes on us moral responsibility.

The important conclusion from this observation is that as long as we scientists are not ready to put our hands in the dirty water of human affairs, as long as we continue to stay isolated in our ivory towers and pronounce weighty statements *deus ex machina* concerning the molecular, the physical laws, biological principles and ethical behaviour, we shall have no influence and our message will carry no weight. Moreover, we shall have to carry the burden of responsibility for the consequence of our own deeds, which we did not succeed in integrating into a decent human way of living.

There is no doubt that much can be done on a long-term basis with our contemporary knowledge. First of all, on the educational level, scientists should collaborate with those directly involved in human behaviour, the anthropologists, the psychologists, the sociologists, and give their help in scientification of the treatment of the human personality problem. Thus, as in the new development which started with Piaget, science and psychology could collaborate in helping the maturation of young people living in different cultures.

Piaget found that during the cognitive development of man, at the age of about 14–15 years, a "formal structure" develops, concepts acquire meaning and moral dicta find their justification. It is this structure which has to grow and form in children, since without it the person remains immature. The maturation of man, and his readiness to take social and ethical responsibility, are the primary problems of education.

But even more important, because the formal structures of our contemporaries are insufficient to embrace global problems, a jump is required from the present-day structure to higher order, internally directed, structures. I imagine Prof. Prigogine will agree that I may call these cognitive structures "dissipative structures of the human mind". Dissipative structures are based on flow processes which by their interaction organize matter and are capable of passing from one degree of organization to a higher degree of organization.

I believe that the informational flows of the mind interact to form organized structures which are capable of passing to higher levels of integration. The studies of Maslow in the States indicate that every man can reach these higher degrees of maturity and with appropriate education can develop the inner formal structures of a responsible personality. Moreover, it was shown that men who have reached high cognitive levels have generally strong artistic interests and are highest in the moral sense. It follows that the maturation of personality to higher levels leads to the evolution of the type of man who could cope with the problems of the present.

But we are pressed for a solution of the immediate problems which cannot wait for the maturation of mankind. The rate of development of the scientific processes is so rapid and the technological consequences of science so urgent that we cannot leave for posterity the solution of our problems.

Can scientists do something about it? I believe we can, and should. Science is now an enormous social enterprise. The percentage of youngsters in the more developed countries who continue their studies to university level is very high. In the United States, there are 7 million students in universities or schools of higher learning; the Russians claim that about 20 million youngsters, that is about 80% of all the youth, attend schools of higher learning.

But science is still based on a classical pattern of a hierarchial, meritocratic society which does not fit present needs. It is based on the principle of the catholic church, busy with its cardinals, its bishops and the promotion of its priests. It is still assumed that science is the privilege of an international elite.

Although the ivory tower attitude will remain of importance for scientific research, scientific education is no more a matter for the elite, but is a major social enterprise. Hence the morality of science and its ethical principles have a direct impact on modern society. It is up to the scientists to undertake, now, the adaptation of themselves to the needs of contemporary mankind.

If there will arise a group of pioneering scientists, who can show how to humanize science and help in synthesizing human needs and scientific development, science will be able to survive, and help mankind to move towards the "promised land". It is the land which for many centuries was the dream of the humanists, the philosophers and the enlightened scientists — the haven which was described by the great Bengali poet — Rabindranath Tagore — as

"Where the mind is without fear and the head is held high.
Where knowledge is free.
Where the world has not been broken into fragments by narrow
  domestic walls.
Where words come out from the depth of truth.
Where tireless striving stretches its arms towards perfection.
Where the clear stream of reason has not lost its way in the dreary
  desert sand of dead habit.
Where the mind is led forward by thee, into ever widening thought and
  action.
Into that heaven of freedom, my father — let my country advance."

Discussion

Prof. Manfred Eigen, Göttingen

It is very difficult to initiate a discussion on this subject because it is so easy to say everything about nothing, or nothing about everything. This morning we have been seeing how much we can learn from looking at biology, and I think we might also ask the question how much cannot be learned by looking at this discipline.

Now let me come to the first point, that there is no evolution without trial and error. In the process of biology it starts at the molecular level. Such systems are quasi-stationary and will always be unstable unless they grow. But there are two ways of growing. One is to grow in quantity, and the other is to grow in quality.

We realize that we cannot go on growing in quantity, since population has almost reached the ceiling, and it is our problem now to find out how we are to grow in quality. We cannot resolve our problem simply by saying all progress has come to an end. I would say that that would also be the end of society. Society is unstable because we are making errors, and errors mean fluctuation; such a society will inevitably be unstable.

Now my second point is that if we reflect on society there is an element which we cannot learn from biology. That was very clearly said, I think, by Aharon Katchalsky in his lecture. If we look at evolution we start with molecules, and we see how molecules start to cooperate to form cells; they do not care about the single molecule, the molecule will be synthesized, will decompose and it is of no importance what a single molecule does once a higher organizational form has been reached.

Now cells start to cooperate into organisms. We ourselves are organisms and we do not care a bit about the single cell of our organism. But if we go on and say that cells finally evolve into humans, and humans start to cooperate and form societies — we cannot say we do not care what a single individual does. There is a big step, the step to a self-conscious, human individuum, that is much bigger than the step from the molecule to the first living cell. We cannot learn about this step simply by reflecting on evolution.

It is a completely new event in evolution, and we have to study it in order to learn what to do, because we cannot go on and say that the society is the superior part, once the individual's self-consciousness has come about. There is certainly much in what Fritz Cramer has referred to as a type of asceticism that we have to learn about, but it cannot be an unlimited asceticism just in favour of society. We have to preserve the individual and not, in developing society, kill his individuality.

So we come back to our first dilemma, and the difficult point in opening such a discussion is to avoid staying on this general level, and to find out what are the single problems. What I would suggest is that we do not go on talking in general terms but try to specify what are the immediate problems, and what science can do toward solving them.

Prof. Reinhard Bendix, University of California, Berkeley

I am one of the few social scientists in this gathering, and perhaps it is not entirely inappropriate towards the beginning of our meeting for a social scientist to speak.

Let me be very brief. One speaker spoke of asceticism, the other of modesty. Perhaps that could be symbolised by a reversal of the title of this conference. Is it inappropriate to speak of the impact of society on science? Let me specify. When you speak of asceticism, one form presumably would be not to do what you are inclined to do. Since you are biologists, what, I would ask, would be your attitude towards a moratorium on research in a given field?

I happen to know that in some fields of biology this is possible and, indeed, specifically discussed. In some types of organ transplants the decision was made that certain types of surgical techniques should be discontinued because the fatality rates were regarded as too high. However, this type of moratorium concerns an applied field. One of the questions which I would like to pose, is: What would a moratorium of this kind mean in fields of fundamental research, where tests of success or failure are not immediately visible? It seems to me that especially in fundamental research in biology, with the problems of genetic control that both speakers referred to, it becomes very difficult to say what will be the consequences of a given innovation.

And let me remind you that the scientific enterprise is based on a kind of *laissez-faire* within the community of experts trying to develop promising lines of research precisely by excluding considerations of the consequences that have in the past been left to others to evaluate. I would like to ask the scientists here how they propose to address themselves to this question of political decision-making related to scientific progress.

Discussion

Chairman: Prof. John C. Kendrew, Cambridge, England

I think these are very pertinent and difficult questions. I seem to remember that Rutherford around 1933 was asked whether there were going to be practical consequences from the work of the Cavendish Laboratory in nuclear physics. He said he thought it was ridiculous that this research would have practical consequences. Rutherford was a very great physicist, but he got it completely wrong. If we are asked to predict the consequences of the research we are doing, how are we going to make better prognostications about the consequences of this research than Rutherford did? This is the kind of problem which faces us.

Prof. Chaim L. Pekeris, Rehovot

While we are on this question of the moratorium on research, since we have gone so far back as Rutherford, I wish to go a little further and point out that it is not a new idea. In 1776 Newton wrote to the Secretary of the Royal Society, Oldenburg, recommending that a moratorium be put on the researches which were being conducted by Boyle, the author of Boyle's Law.

Newton was particularly infuriated by an article which appeared a month before in the Philosophical Transactions of the Royal Society by Boyle entitled *The Incalescence of Quicksilver with Gold*. Now most of you have blank faces when I use the word "incalescence" except one man, Prof. Scholem. This is a subject in alchemy, and what Boyle published in 1776 was a new recipe for making mercury react with gold and develop heat, and that is "incalescence".

Newton, in one of the very few cases where he referred to other people's work, took the trouble to write to the Secretary of the Royal Society, asking him to intervene with Boyle to put a moratorium on researches on incalescence of quicksilver with gold because of the great menace such research would hold for the future of society. Now, closer examination of the motives that prompted Newton to write this letter turned out to be not so altruistic. Actually Newton was afraid that Boyle had the jump on him in discovering the philosopher's stone, as he, Newton, was heavily engaged in alchemical research. So people who are now proposing a moratorium on research should not in any sense feel they are suggesting something new.

Prof. Albert B. Sabin, Rehovot

Since Prof. Pekeris has indicated that the question of moratoria is not new, it is well to remember the subject of this symposium is also not new. There is always a new angle, however, and we see the biologists' dreams are the humanists' nightmares. When a biologist begins to dream about the things he may possibly control, it's enough to give one the shivers. But the question is what to do about it. Shall we turn our attention to gaining the wisdom with which to utilize the knowledge, or shall we have a moratorium?

Prof. Sir James Lighthill, Cambridge, England

As a non-biologist, I thought I would mention some areas which it seemed to me biologists might have emphasised. Prof. Cramer pointed

Discussion

out that the human genes have not evolved since the Early Stone Age, and suggested that man had brought the evolution of other species to an end as well; but it seems that since the Stone Age domestication of plants and animals has constituted an acceleration of the process of evolution. This could be one of the areas where we have most to contribute in alleviating the scarcity that still prevails in the greater part of the globe.

In discussing the evolution of *Homo sapiens,* Prof. Cramer seemed to have emphasized intelligence at the expense of the other characteristics which also seem to single out the species. One is the long period of parental dependence leading to a more fully developed altruism in the species capable of the sort of extension that Prof. Katchalsky was talking about. Surely the problem is, as Prof. Katchalsky said, the extension of the altruism that the individual human being feels to the whole global species. This is an area where science and technology have a very positive contribution to make.

One recognises that transport technology has enormously altered the extent to which it is possible to extend the altruism that originates from the feelings in the family or tribe to the globe as a whole. It has been very striking to witness the effect jet aircraft have had in the last twenty years. From the point of view of the impact of science on society the jet aircraft is often regarded as an anti-human device. On the other hand the transformation of the world into a neighbourhood by aircraft, which has not yet reached its conclusion by any means, may be one of the areas where technology is making possible man's greater humanity, because it makes it easier to love one's neighbour on the other side of the globe.

I just want to add one thing about moratoria because we have had moratoria on pure science occasionally within our living memory. There was such a moratorium during the Second World War when the pure scientists worked on the application of science to war. This led to a vast improvement in the level of technology and as a result much better instruments were available when science revived after the war. As a consequence pure science was soon at a much higher level than it would have been had peace prevailed.

It is, I think, very much a matter for consideration whether a moratorium on pure science on an internationally agreed level might not be seriously considered. It would of course be most important that there should be a generally agreed feeling of urgency about the situation. At the same time the population explosion and the problems of the underdeveloped countries do give goals that could, in principle, unite scientists. It would be very interesting to see if a short, perhaps three or four years' global attack by pure scientists on the real problems of scarcity and overpopulation might not produce quite remarkable results. As I said earlier, it seems most unlikely that pure science itself would suffer from a temporary deflection of the intellectual energies of the scientists into other directions.

Prof. Victor F. Weisskopf, M.I.T., Cambridge, Mass.

I would like to say a few words about a moratorium in science. The example the speaker before me has made is a very interesting one, namely the Second World War, but I think one ought to interpret it, to my mind, in a different way.

Science, in particular pure science, is not only production of knowledge, it is also the production of people. And these people, the ones that are trained in research, are very necessary to solve the many problems facing us. Now we were lucky in the Western world that World War II lasted only four years. If the moratorium it produced had lasted longer there would have been a lack of young scientists. Germany is only now recovering from the suppression by the Nazis of many branches of pure science.

The problems we are facing will need generations to solve, and if we destroy our basic science establishment in order to solve those problems we shall not have the flux of people, the spirit, the attitude, for the task. This is why I think a closer analysis would show a moratorium on science would make it much harder to solve the problems we are facing. And I mean this not only in the severely practical sense that no new knowledge will be found, because I find this the less important thing; but the spirit, the attitude, the tradition of basic science is the one thing that we will need more than ever.

Prof. W. Van Cotthem, University of Ghent, Belgium

We heard Prof. Katchalsky say that one of the important tasks of science is education of society and especially of the youngsters in the society. At the University of Ghent we are on our way to split up the education of scientists into two different branches, one branch designed to train research workers, and the other for scientists intending to teach in the schools. Now I would like to ask Prof. Katchalsky if we should not think about a special training of scientists for out-of-school education especially in underdeveloped countries.

Prof. Carolina MacGillavry, Amsterdam

I agree with Prof. Katchalsky that one of the most important things we can do is to teach our students the value of facts, the *"Realitätsprinzip"*. At the same time I believe it makes us vulnerable and in social affairs often incredibly naïve. How does this come about? I think in this way: because we base our judgments on the value of facts, we can rightly be critical of statements by colleagues in our own discipline.

We tend to accept what scientists in other disciplines say because we assume that, like us, they are scientists having respect for the facts. But what of politics? It has become the fashion for scientists to sign all sorts of statements and manifestos without knowing the facts behind them. This is an example of the "modesty" of scientists; they are modest, but they tend to believe their name means more when used in this way than that of someone who really knows what it is all about. For myself, the greatest difficulty is the finding of my role in society.

Prof. Jeanne Hersch, University of Geneva

I should like to tell Prof. Katchalsky that I am especially pleased with the philosophical part of what he said.

I think that if science has played a harmful role in relation to the meaning of human life, it is because it has practised, and I believe still does practise to a large extent, a research activity which tends to reduce meaning and to substitute for the meaning-of-meaning an explanation which takes away the meaning of the thing studied.

It is a little as if for the purpose of studying a living thing, one cuts it up into little bits and then finds only death. In the same way one explains meaning by taking it apart. And that is why I think that the remedy should not be sought in a moratorium, but in a conversion of science to the notion of limitation.

You yourself have posed the question, "Should we do such and such a thing because it is possible?" And I think there are cases now when we must say "No". Thus the idea of limitation. For example, even if moral laws had, in part or in general, a utilitarian origin, that does not alter the fact that the way they impose themselves on us today is not utilitarian, since our actions may be guided by moral values rather than profit and utility.

So we see another factor has entered the picture; it is that element which makes the law of morality impose itself. That is why I think also that the meaning-of-meaning of science must be rediscovered. Actually, when one speaks of the basic foundations of science, one is not referring to the meaning of science, but the meaning-of-meaning of science.

Prof. F. Cramer, Göttingen

I thank the audience for many interesting remarks. I can only discuss a few of these and in the first place make a few comments about what Manfred Eigen said.

Evolution is an unstable system based on trial and error and one can, as he did, demonstrate mathematically that such a system must propagate, must perform in a way which I called "escape forward". Now this escape forward has reached a quantitative ceiling because, at least in our planetary system, material resources are limited.

But it is possible that certain non-material developments occur, for instance through brain research, which give us a possibility of further propagation, of further development. The question today is whether we have time enough to wait for this new qualitative step which I, in a simplified way, have called the eating for the second time from the tree of knowledge, because material evolution, the development of technology is developing so rapidly.

And here I might insert a remark on the moratorium. The moratorium of the Second World War was, in my opinion, very unfortunate because it shifted interest towards application. As a result application outstripped our understanding of the consequences, and developed much faster than basic research. We are now faced with an overwhelming application of technology but with a great lack of basic knowledge and basic consciousness about its application.

The question is, how can we fill the time-gap between further development of our consciousness and the present high state of technology. I think the challenge which we have to meet as basic scientists is that of spreading our knowledge in order to create a common consciousness in the advanced societies, and later of course also in the other societies. We must develop new moral laws about the application of science.

To give you a simplified example, in my opinion it is immoral to give a person, let's say penicillin, who does not know how penicillin works and what its consequences are; it is as immoral as giving a negro in Central Africa glass beads in exchange for his cattle. This is an over simplification, but what I mean is that society has to be brought to a level of the understanding of science so that it can have an awareness of the consequences of science. Only from this level might we be able to bear the consequences of science.

Discussion

Prof. Aharon Katzir-Katchalsky, Rehovot

May I start answering the remarks or the questions not in the order that they were posed. First of all, a few words on science for developing countries. As is well known, the United Nations and some of the developed countries have invested billions of dollars in science in the underdeveloped countries, with very poor results. The general recognition is that most of the money went into the wrong pockets and had very little effect on raising the standard of living and the intellectual standards to permit the people to use science.

It was very soon discovered in underdeveloped countries, and to a certain extent in the underdeveloped communities in Israel, that there is no difficulty in teaching the underdeveloped people techniques. They can run a car, build and operate a factory, but they cannot create in a scientific way. And generally it is found that at a certain age group the drop-out is considerable; they cannot follow the conceptual structure of a scientific society.

When you make a deeper analysis you discover that a lot of the conceptual foundations of Western society are imbued in children at a very early age. Causality, numerization of phenomena, evolutionary concepts, irreversibility, all these are not the product of the school. They are the product of the home, and the social atmosphere. They are lacking in societies that do not know how to measure, how to translate phenomena into numbers, whose causality is mythological, who do not recognize evolutionary trends in the life surrounding them.

So from this point of view the major problem of education — and this is where scientists and educationalists have to collaborate — is to find out what are the necessary foundations that can be isolated by the scientist himself. Then we must translate these foundations into a language that can be incorporated more or less painlessly into the cultural structure of underdeveloped people. There is no difficulty in destroying their cultures; this was done for hundreds of years.

But to bring culture, in a humane way, and let it become a part of their life, this is the major issue in trying to bridge the enormous gap between the developed and the underdeveloped countries. This is a major problem of modern mankind in general.

Another couple of remarks. I accept Manfred Eigen's comment that evolution is based on trial and error. Eigen himself has developed a beautiful theory on how this process works even on the molecular level, and it is surely one of the major factors of animal evolution. But in human evolution there is an additional factor. The trial and error that continue assume a symbolic form, and the structure which develops in mankind is a symbolic structure.

We can operate and make trials with symbols, and constitute a Gestalt in which we incorporate the results of our *Gedankenexperimente*. From this point of view the trial and error process is not good enough for mankind, because it has passed into a new stage, and it is this stage which for us is of such great importance.

Now about the moratorium. First of all, nobody knows how to stop research, even if such a decision is made. And surely nobody knows how to keep truth under control and stop its dissemination. In Bertold Brecht's *Galileo* the Great Inquisitor pleads with Galileo and asks him to put a moratorium on his astronomical discoveries. The pleading of the Inquisitor has great human significance. He tells Galileo: "With your little telescope you have torn down the skies and instead of skies you have now an empty space. I have no place left to put God and the Angels..."

And he says more: "The central position of the globe which gave so much meaning to human life is undermined. Have pity on mankind

Discussion

and keep it quiet." And the answer of Galileo, a cruel, horrible answer, is the classical answer of science; he has full sympathy with the Inquisitor, but you cannot keep truth in hiding. If he suppresses it, somebody else will disseminate it.

Scientific discovery is the truth of Nature, and you cannot hide Nature. You can keep a scientist in prison, but you cannot keep truth in prison. So from this point of view I am fully in agreement with the remark of Weisskopf that science is not only a road to technology, not only beneficial or horrible to mankind, but science primarily is the great venture of man, the venture of understanding Nature, the great spiritual adventure, and to stop science is essentially impossible.

I am very much for the idea of Mme. Hersch that the meaning-of-meaning is of the greatest importance in the personal development of man, and one of the greatest possible contributions of science as long as science carries meaning-of-meaning. I was very impressed by the summary of the work of Jean Piaget and of Jerome Brunner and other workers, that with the growth of the internal structures in the human psyche everything fits, and a concept which is empty gets its meaning by the formal structure.

This formal structure is nonscientific—and I fully agree that the formal structure of man is the human condition. It is the formal structure which is the road to the highest ideal of the Greeks. On the gates of the Pitia of Delphi was written *Gnothi seauton*—know thyself. You can know yourself only by the existence of a formal structure. But there is no doubt that it is not good enough to recognize the formal structure. Man is developing all the time and the major problem which the scientist is facing is his integration with the humane needs towards the development of higher structures dictated by present day conditions.

Professor Hendrik B. G. Casimir has been a director of the Philips Research Laboratories at Eindhoven since 1946 and a member of the Board of Management of the Philips Company since 1957.

Professor Casimir was born in The Hague on July 15, 1909, and studied theoretical physics at the University of Leyden, in Copenhagen under Niels Bohr and at Zürich with W. Pauli. He has published many papers on theoretical physics, applied mathematics and low temperature physics. In his present capacity he supervises the research activities of all Philips Laboratories throughout the world.

Afternoon Session

Monday, June 28, 1971

Chairman:

Prof. Hendrik B. G. Casimir, University of Leyden

Professor Léon Van Hove is a physicist in the Department of Theoretical Physics at CERN in Geneva. A native of Belgium, he was born in Brussels on February 10, 1924 and studied at Brussels University where he received his Agregation de l'Enseignment Superieur (Ph.D.) in 1951. He was a member of the Institute for Advanced Studies at Princeton 1949–50 and 1951–54, and Professor of Theoretical Physics and Director of the Institute of Theoretical Physics at Utrecht University 1954–61.

Physical Science in Relation to Human Thought and Action

Prof. Léon Van Hove, CERN, Geneva

I was asked to present in thirty minutes a physicist's views on the subject of this symposium, *The Impact of Science on Society*. It is a great honour, but perhaps not such a great pleasure, because I feel myself a little like a child in front of a huge cake who is invited to cut a small slice out of it, and is a little afraid that the inside of the cake is not entirely solid.

Fortunately the very first letter that I received from Dr. Sabin did not mention this subject as the one to be treated and more specifically referred to the philosophical and practical impact of scientific endeavour. That allowed me to think in a more optimistic way of my task, before the real title of the symposium was known to me, and this undoubtedly facilitated my choice of approach.

I will give you a down-to-earth lecture. I will refer to a few historical aspects in the development of physical science. I will occasionally draw a short lesson; I will make one, and only one remark on the future, and I will leave the more lofty considerations and speculations to the discussion.

Physical science as we know it today was born in the seventeenth century. The great founder of the first physical theory is Isaac Newton and what he founded is classical or Newtonian mechanics. This was, and is, an admirable body of knowledge; it was developed and perfected to an astounding degree during the 18th and 19th centuries. It succeeded in describing in very precise terms, by means of a limited number of mathematical laws and equations, a very large class of natural phenomena, e.g. the modes of motion and interaction of bodies like planets, little stones, or complicated machines.

On the basis of Newtonian mechanics scientists were able to describe and calculate the behaviour of a variety of mechanical systems. They could predict invisible properties of these systems on the basis of visible and measurable ones.

There is an almost legendary example of this, which I want to recall, namely the famous episode of the discovery of Neptune. In the beginning of the last century it was realized by astronomers that the orbit of Uranus was not understandable. As a result of purely theoretical calculations by John C. Adams in Cambridge and by U. Leverrier in Paris it was possible to explain the situation by assuming the presence of one additional planet. The very precise calculations of Leverrier predicted where the planet should be in the year 1846; he communicated his findings to John G. Galle, at the Berlin Observatory, and on September 23, 1846, Neptune was found within less than 1° of arc of the predicted position.

This was a great feat, and illustrates, symbolises you might say, why classical mechanics has had such an enormous impact on a variety of things ranging from philosophy to technology. Newtonian mechanics gave us the first mathematically precise formulation of determinism or causality. It gave a precise basis for a mechanistic view of the world. Its impact on other sciences, on philosophy, on thinking in general was tremendous and it deeply coloured our attitude towards technology.

A few decades after this culminating example of the discovery of Neptune things began to happen in physics that profoundly modified the picture. The last hundred years, starting roughly around 1870, have not only seen an explosive expansion of the physical sciences, and an enormous harvest of fundamental discoveries, largely of an unpredicted type, but also they have taught us a sobering lesson.

Newtonian mechanics, despite its spectacular success, despite its admirable mathematical elegance, is quite inadequate, or even worse, plainly wrong, when applied to other classes of phenomena. For example, as soon as one looked at the bodies that are treated in classical mechanics, and concentrated on their properties when they move very fast relative to each other, a hundred thousand kilometers or more per second, it became evident that Newtonian mechanics failed. Newtonian mechanics, in such cases, has to be replaced by relativistic mechanics, as Einstein discovered in 1905. Here the theory is so different that the very basic concepts of time and space have been shattered. From absolute, as Kant wanted them, they became relative, different for different observers.

The new theory, relativistic mechanics, although better than the older one, does not really equal the latter in mathematical compactness and elegance. Of course, the great successes of Newtonian mechanics remain as strong and as brilliant as before, for the class of phenomena where it holds good. It was the theoretical status of Newtonian mechanics which was degraded—it was no more than a low velocity approximation of the new theory of relativistic mechanics.

From then on we knew that any theory can have the same fate. Beautiful, impressive, successful as it may be, it may some day be degraded to be an appendix, an approximation to something else we don't yet know. Here again, of course, the lesson extends beyond the borders of physics. The greatest success of a scientific description of a given class of phenomena offers no guarantee whatsoever that we can trust it even in neighbouring domains. In other words, scientific generalization, scientific extrapolation, involves great risks and requires very critical attitudes.

During the last hundred years physics has also given rise to other types of development that are of profound significance because we can draw on them for many other purposes. These are the developments that took place when matter began to be studied at the atomic level. It was discovered that at this level the deterministic laws Newton launched into the world are no longer sufficient, and that one needs statistical laws to describe and predict the properties of matter.

The first stage, which started about a century ago, is the following. It was clear that the laws of classical mechanics do apply to a good approximation to the atoms and the molecules which are in an ordinary liquid, like water, or in an ordinary solid, like salt. However their application remains essentially fruitless until they are supplemented by statistical considerations which centre around a new concept, the concept of entropy or the statistical measure of disorder. Disorder plays a role, and you have to grasp it mathematically by means of appropriate mathematical concepts.

Then, much later, in fact in the nineteen-twenties, another stage in the study of physics at the atomic level took place. It was the stage at which one went into the internal structure of single atoms and molecules. Here statistical considerations turned out to play a more fundamental role in the sense that a completely novel situation occurs at the level of raw experimentation, that is identical experiments turn out not to give identical results. Determinism, on the level of single experiments, does not hold.

It is only when you repeat a large class of identical experiments that laws of physics can still be enunciated. It is only the statistical distribution of results of a large number of such experiments that is predictable. Determinism has been replaced by statistical causality.

The emergence of statistical causality had of course numerous and far reaching consequences outside of physics, in particular at the philosophical level. In many sciences, and especially the social sciences, it turns out that the central role is played by the use of statistical laws and by the search for forms of statistical causality. Physics gives us precise, instructive, detailed examples, not only of the power and the diversity of what statistical theories can be, but also of their limitations. It may be good to remember these examples when one goes on to apply statistics to much more complex and less understood theories.

As we all know, some of the natural phenomena which are studied in pure research may some day find applications of a practical nature. In physics it so happens that in a short interval of ten years two discoveries were made that found almost immediately technological applications that have had immense impact on human affairs.

Nuclear fission was discovered in 1938–39; it produced nuclear weapons as well as peaceful nuclear power and it took only seven short years from the scientific discovery to Hiroshima.

The transistor was invented in 1948 in the field of semiconductor physics, a branch of solid state physics; again, less than ten years later this device revolutionized a good part of electronics and in a fundamental way the whole field of electronic computing. Still, neither of these discoveries can be called particularly fundamental in itself. But nevertheless the world was profoundly changed, was in fact shaken by the indirect consequences of these discoveries, not only in technology and industry, but in politics.

Of course, most scientific findings have no such direct impact on human affairs, and personally I am happy for it. Even those findings that form essential contributions to fundamental knowledge, for example, some of the deepest aspects of relativity, the relative nature of space and time, have found no technological application.

Another amusing example is the discovery of superconductivity, that incredibly fascinating state of matter at very low temperatures which permits it to maintain electric currents for months without dissipating any heat. This beautiful discovery was made in 1911; it is only now that the most timid applications in technology are beginning to emerge. Still, the lesson is there, and we have to live with it; that is we have no way of telling in advance whether a piece of scientific research will have technological applications or not, and if it does — whether the consequences will be good, bad, or both, since usually it is both. The social, the political, the moral consequences of science have become a concern of all, scientists and non-scientists, and research is now a matter of national and international politics.

But there is of course one more reason why science, and physics in particular, have become a matter of political concern. This is the emergence of big science. Some of the most fundamental fields of research, such as nuclear and high energy physics, as well as astrophysics, require very large instruments. Here the scope and the cost are colossal. Centralization and planning have become a necessity in these big science activities.

They gave birth in the fifties to new types of research institutions, very large laboratories, either national or international, devoted to a single branch of pure science. Some of these institutions have become very powerful bodies. I shall not say much about them because, after all, they defend their interests very well, and they have developed the great art of presenting their achievements, which in some cases are considerable. What I shall do in the short time available is to raise one aspect of this development which perhaps still calls for some improvement, and try and give an example. I shall try to make a case for the thesis that big science has not yet reached its logical conclusion at least in the organizational plane. I shall illustrate my point with high energy physics, where I happen to have worked for the last ten years, and which is characterized by extremely large accelerators, synchrotrons and other machines of the same type.

The largest of such machines now operating are the Russian 70GeV proton synchrotron, and the West European proton synchrotron at CERN in Geneva. A very much larger synchrotron is nearing completion in the United States near Chicago, and ten West European countries have decided recently to launch the construction of a similar machine at CERN.

But let us now step back and look around on a world-wide basis; let us list the most competent, gifted and creative groups of high energy physicists. If we do this of course we shall find a few of these very, very gifted groups in the countries where the accelerators are located. But we also find, for example, two of the best experimental high energy groups in the cities of Warsaw and Cracow, while perhaps in some way the most brilliant and most original of all theoretical groups is that based at the Weizmann Institute in Rehovot.

These groups have no official access to the most advanced accelerators I listed a few moments ago. And the reason is very simple, although unfortunate; they happen not to belong to the nations which own and operate these large machines. Of course, Polish experimentalists, Israeli theorists, as well as Polish theorists and Israeli experimentalists are regular visitors at these big laboratories. But they are, nevertheless, in some way a special class of visitor, and to call things by their name, I mean second-class visitors.

And the situation is not entirely deprived of a slight, but bitter taste of scientific charity. In particular, one point should be stressed: *advice*. These people, with their exceptional competence, are not used where they are needed most, that is in discussing and deciding the difficult questions of scientific policy on which so much depends for the success or failure of large research projects. It is here, I believe, that big science has to take one more step if the best use is to be made of its very large investment.

Some day, and the sooner of course the better, the policy must be that all these very best brains on a world-wide basis shall be involved in all stages of planning, constructing and exploiting the most advanced research instruments. This is a challenge for the future; it is a worthy one and I think it points the way to a change for the good in big science, especially in what it means for non-scientists.

Indeed, let us recall the image that was evoked a generation or more ago in physics when reference was made to big science. Well, to the physicists it was the image of gates and complicated passes, in fact the image of secrecy, weaponry, cold war. The situation at present regarding big science, and even more so, in the future, is more encouraging, namely in the direction of world-wide co-operation on the very giant instruments which are needed if we want to pursue the very long quest for fundamental knowledge.

Let me now end by summarizing the three points I have tried to make. Firstly, the development of physics has shown a mathematical description of natural phenomena is possible, and that it provides the basis for theories with predictive power. The concepts of causality, of statistical law, have found precise meanings in physics, and have been confronted with extensive classes of experiments. Their development in that particular science can serve as a model, and also as a lesson for the application of mathematical techniques in other sciences. It illustrates both the power as well as the limitations of such techniques. All this clearly implies a lot for thinking, scientific knowledge and philosophy, as well as action, prediction, planning, and technology.

Secondly, physics discovered and studied the properties of natural forces and of material systems; it provided, occasionally, ways of controlling some, or even utilizing some of them. This has had consequences of a non-scientific nature which have had a tremendous impact on human affairs. Science is now a matter of politics, national and international.

Thirdly, the present trends in fundamental scientific research, particularly in physics and astrophysics, lead to the investigation of ever-more remote classes of natural phenomena which need ever-more powerful instrumentation. This development will require the formulation of a global science policy by which the world's most advanced research tools will be planned, constructed and exploited in a concerted way. This must be done with full participation of all the very best brains available on a world-wide basis. And the reason is not only one of general international morality, but also a much simpler one; and that is that the total number of the very best brains is rather small. Their integrated wisdom is certainly not too large for the difficult tasks which lie ahead on the long road towards ever-deeper layers of fundamental knowledge.

Professor Chaim L. Pekeris is Head of the Department of Applied Mathematics of the Weizmann Institute of Science. Born June 15, 1908 in Alytus, Lithuania, he came to Israel in 1948.

Professor Pekeris studied at the Massachusetts Institute of Technology, receiving his Ph.D. there in 1934. From 1946–68 he was a member of the Institute for Advanced Studies at Princeton, and from 1937 to 1941 he was a member of the Faculty of the Massachusetts Institute of Technology. Professor Pekeris has done research in fluid dynamics, geophysics and atomic physics. He was awarded the Rothschild Prize in Mathematics in 1966.

The Impact of Physical Sciences on Society

Prof. Chaim L. Pekeris, The Weizmann Institute of Science, Rehovot

First, allow me to introduce our group. We physical scientists are known to be among the safest drivers in the world. The few accidents that do occur happen mainly when we back up and drive in reverse. The reason for this is that we are so polarized in the forward direction that we have lost the ability to look backward. I have therefore forced myself to examine the past with a view to ascertaining whether recent experiences with projects in the physical sciences may already contain lessons bearing on the theme of this conference.

Now I do not intend to deliver an apology on behalf of the physical sciences or to extol their virtues. I rather wish to discuss the situation in terms of our own standards. Two controlling factors have emerged from the Los Alamos experiment of 1942–45:

1. We have demonstrated that we can develop new technologies starting from an initial state in which these technologies did not exist.
2. We have shown that in wartime we can organize ourselves into a coherent team which works with extraordinary enthusiasm and efficiency.

These two factors present us physicists with a moral problem: Why should we not initiate in peacetime a project or projects which will be of direct benefit to society?

Admittedly, we cannot expect to reach the level of efficiency and concentration that was achieved in Los Alamos, but some fraction of that is not unreasonable to expect. With this in mind I wish to examine some recent projects in the physical sciences which are potentially of direct benefit to society. I shall try to evaluate their mode of origin and operation and to determine how successful they were, or are, in achieving their technical goals.

We shall start with a project of discrimination. The problem is how to discriminate between a seismic record stemming from a nuclear explosion and one from a natural earthquake. Since the Partial Test Ban Treaty was signed in 1963, outlawing tests of nuclear weapons above ground, no agreement has been reached on the means of verifying the observance of a ban on tests below ground. In 1963 the technological problem of identifying the signature of a nuclear explosion was an open one. A research programme sprang up in both camps, East and West, with a surprising degree of exchange of information, with the result that today no underground nuclear explosion bigger than about 2 kilotons can pass undetected.

This threshold was 10 times bigger only three years ago. Every year the researchers in the field gather for an international conference where information is exchanged. In 1969 the Russian delegation was taken on a tour of the seismic array in Norway (NORSAR), and last year they visited a similar site in Sweden. The participants are mainly from academic institutions, and they have exhibited a degree of inventiveness, both experimentally and theoretically, that is impressive.

The scale of operations is a modest one: about a hundred million dollars. The project has been successful in attracting new talents, the leadership at the top is adequate for the needs, and the results to date are rewarding. I consider this a successful venture on an international scale, in spite of the background of the conflicting interests of the two camps, and the sensitivity of the subject. Today we possess a body of knowledge about the nature of earthquakes and explosions together with expertise, which did not exist seven years ago.

After the Los Alamos experiment, the late Professor John von Neumann embarked on an ambitious programme aimed at revolutionizing the physical sciences through the use of electronic computers. He designed and built the Johniac computer in Princeton, and the first target of application was numerical weather forecasting. The idea was to start with a knowledge of the initial state of the atmosphere and to let the computer integrate the dynamical equations to produce mathematically the state of the atmosphere in the future. Initial feasibility trials were carried out by a Princeton team headed by Jule Charney, now at M.I.T.

Progress at first looked promising. When I saw Neumann last at the Walter Reed Hospital in Washington in 1955, he was in high spirits, claiming that his team had licked the short-range forecasting problem—for a few days in advance—and that the time had come to plan the attack on the long-range weather forecasting problem. He immediately proceeded to figure out the size and speed of a computer that would be needed for long-range forecasting. Unfortunately, Neumann died shortly thereafter, the Institute of Advanced Study chased out the whole Neumann team, and Charney moved to M.I.T. Work has continued and is going on today, but at no time were the existing computers adequate for the needs. The problem is to introduce finer details in the assumed model of the atmosphere and still have the computation finished ahead of the weather change.

The status today is as follows:
1. It has been estimated that under ideal conditions (i.e. when the initial state is measured in maximum of detail and the computations

are carried out at the highest accuracy), it should be possible to predict the weather for a maximum of about 30 days.
2. A world-wide organization (GARP) was established to gather weather data, including photographs from satellites.
3. As of 1968, weather data from both hemispheres are used in experiments of numerical forecasting.

I got the impression from a review of the recent literature that while there is abundant solid talent in the several teams working on this project, there is no leadership at the top such as was provided by Neumann. The project requires expertise in numerical analysis, fluid dynamics, outdoor type of instrumentation, and data-processing, coupled with a healthy dose of common sense. It is not excluded that the governing mechanism of the weather will turn out to be actually simpler than it is thought to be now, and that a new approach may be called for. This is the class of project which owes its initial success to a strong personality, but which also carries the danger of being stalled if the driving personality disappears.

In the evening of March 27, 1964, when the city fathers of Anchorage, Alaska, were holding a ceremonial meeting on the occasion of Good Friday, an earthquake struck the city which caused over 100 deaths and completely wiped out the commercial fleet in the harbour. The councillors ran for cover (except for the town clerk, who did not budge until he had recorded that "the meeting was adjourned, on the motion of city hall"). Thereupon, President Johnson called in his chief scientist, Dr. D.F. Hornig, and said, "Your next assignment is the prediction of earthquakes." So Hornig called in his advisers and, surprisingly, a group headed by Dr. Frank Press actually agreed to pursue this subject.

Surprisingly, because until the Alaskan earthquake, the topic of earthquake prediction elicited only ridicule among the professionals.

Today, after 7 years' operation of the Press Panel, a substantial body of knowledge has been acquired as to pre-earthquake phenomena. While admittedly no successful earthquake predictions have yet been made — and this holds for the San Fernando (Los Angeles) earthquake of February 9, 1971 — we seem to be getting there. How to act once we get there is another psychological problem which need not concern us at the moment.

The advent of the Press Panel coincided with a new phenomenon that appeared on the seismic scene: earthquakes induced by the injection of fluids underground. At Denver, Colorado, a disposal well was sunk to a depth of about 3.7 km for the purpose of injecting waste fluids. The fluid injection was followed by a series of earthquakes. They are believed to have been caused by the relief of friction between the faces of faults where stress was accumulating.

Once the reality of the phenomenon proved convincing, the idea arose of deliberately using fluid injection to release the accumulation of stress in fault areas which are known to be foci of earthquakes (such as the San Andreas fault in California). Indeed, experiments of this nature are being carried out in California, not without some trepidation: it is not guaranteed that the artificially induced earthquake will be a "small one", and no geophysicist wants to carry the guilt of engineering a devastating earthquake. In Japan, earthquake prediction is being pursued on an even more intense scale.

This project is of interest, first because of its mode of origin: the White House decided that it had had enough of earthquakes and ordered the court magician to get rid of them; secondly, because it illustrates that a subject which only a few years ago was relegated to mystics and fortune-tellers has proved, after the geophysicists were induced to give it some thought, to be amenable to rational investigation. The grim significance of the subject is that, based on past statistics, we may expect in the United States alone a very severe earthquake, on the San Francisco or Alaska scale, before the year 2000.

I believe that physicists can contribute to society's welfare if they will collectively take an interest in this ongoing project of earthquake prediction by occasionally injecting new ideas and approaches. We should do it not because of an outstanding debt to society, but because we owe it to ourselves by the dictates of our own standards.

My thesis is that we can contribute to society's welfare by taking the initiative and the responsibility for some project which can be of direct benefit to society. The alternative, it seems to me, is for us to have to spend much of our time in the passive role of advisers to public bodies, sitting on panels in the role of the unprejudiced, but in many cases also uninformed "experts", and thereby exposing the community of physicists to implications over which it has no control.

As an example of the impact of science on society in the raw, I wish to cite the recent crisis in the U.S.A. with the supersonic transport, the SST. This is the companion to the Concorde and the Russian TU-133 supersonic planes. I managed to ignore this subject, along with others of this nature, until I visited the U.S.A. at the end of March, 1971, when I was told, by a colleague who became involved in this imbroglio, that the SST was voted down in the Senate because Professor James McDonald from Tucson, Arizona, claimed that the SST will cause an increase of some 10,000 cases of skin cancer a year. This cancer, according to McDonald, would be caused by what appeared to me at first a bizarre physical process.

It is claimed that (1) the SST would increase the water vapour content of the stratosphere; (2) this would cause a depletion of the ozone in the stratosphere, which in turn (3) would result in opening an ultra-violet window in the ozone absorption region of 2900 Å, and (4) hence the 15% increase in skin cancer incidence. When I protested that as a referee I would not accept such a theory for publication in a geophysical journal without further work, I was told that in the case of a possible health hazard the criterion is not whether a process has been proved, but whether one can disprove it for certain. As long as you cannot disprove it, you cannot disregard it.

Precise information on the ozone layer in the stratosphere was already available in the thirties, due primarily to the efforts of the Oxonian team of Lindemann and Dobson. Lindemann is perhaps better known to you as Lord Cherwell, Churchill's science adviser during the war—and atmospheric ozone was his specialty. The stratosphere has no garbage disposal system; radioactive debris has been observed to persist there for periods of over two years. Such experts in this field as Professor Newell from M.I.T. have therefore warned Congressional committees of the danger of pollution of the stratosphere by the SST.

The U.S. Department of Transportation has appointed a committee, called the SST Environmental Advisory Committee, to investigate the problem. In January, 1971, Professor Fred Singer, an atmospheric scientist, formerly from the Universities of Maryland and Miami, was appointed chairman of this Committee, after the first chairman, Dr. Tribus, resigned. Dr. Singer was called before a Senate Hearing Committee early in March, 1971, prior to the vote of the Senate on the SST (which was scheduled for the end of March). The trend of Professor Singer's statement[1] was that, while his Committee had not yet finished its investigations, and would not come up with a report until some time in the future, he felt that the opponents of the SST had assumed the worst possible contingencies in every factor.

He cited a series of factors, aside from the SST, which are of considerable potency; among them, for instance, the increase of water vapour in the stratosphere due to methane. One of the sources of methane in the atmosphere is flatulent cows. He asserted that, while no definite conclusion can be reached until all the data are gathered and analyzed by his panel, "our best present judgment is that the effects of an SST fleet on the stratosphere composition will be within the normal range of variability."[2] The last paragraph in his deposited statement reads, "In the balance, I believe that the question of whether we should or should not have an SST must be decided on the basis of economics, with the environmental effects having a very small weight indeed. If the SST is to be turned down, let's not turn it down for the wrong reason."[3]

This concluding paragraph elicited a furious denouncement by Senator Proxmire from Wisconsin[4]: "I am chilled, shocked and outraged by the way you approach your problem. You have given us the conclusion before the study has even begun.... This is like a fight with my wife in which her mother is the referee.... And you are the man on whom many Senators will rely." When Mr. Magruder from the Department of Transportation sprang to the defense of Dr. Singer, Senator Proxmire accused him of packing the Singer Panel with pro-SST men. Senator Proxmire then read off a list of twelve scientists who, in his opinion, were deliberately kept off the Panel. He asked[5] why Magruder didn't appoint Lord Snow, noted English author and scientist, and Dr. George Wald, Harvard biologist, to name just two.

I am citing these sordid details in order to bring home the nature of a real impact of science on society. Clearly, Professor Singer's statement, in which he cites his personal preliminary conclusion, would have been appropriate at such a forum as an American Physical Society meeting, but most unsuitable at a Senate hearing where practical people have to make up their minds a few days before a scheduled vote.

Now this is not the end of the story. At these same hearings, Admiral Rickover testified that, although in 1963 he warned about environmental dangers, he believes now that the pendulum has swung the other way. He is opposed to those who are promoting an anti-technological scare. He stated that even steam locomotives were opposed by anti-technologists. The Royal Society in England[6] "warned that at speeds over 30 miles per hour the air supply to passenger compartments would be cut off and passengers would die of asphyxiation." (By the way, I have not succeeded in tracing the source of this warning, either from the Admiral or from the Royal Society of London.)

"The College of Physicians in Munich warned that at these speeds (i.e., 30 miles per hour) people would lose their sight because of the blurring of the surroundings..." Thomas Edison waged a campaign against the use of A.C. versus D.C. in electric power transmission. To demonstrate his point, his supporters were alleged to have staged public electrocutions of dogs. One effect of this propaganda was the adoption of the electric chair by New York prisons, which I presume operates on A.C.

But the story continues. On March 16, Professor Charney of M.I.T. was approached by the scientific director of the Boeing SST programme, who asked that Charney lend his name to a statement in support of the

SST, a statement being circulated by the President's Science Advisor himself. Professor Charney, an ardent opponent of the SST, promptly reported this to Senator Proxmire, and a press conference was held in Washington on March 18. Present there was also Professor James McDonald, the originator of the skin cancer scare.

The pro-SST camp managed to weaken the weight of Professor McDonald's testimony by drawing from him an admission that he had previously testified before a House Committee, where he warned that the power failures in the state of New York may have been due to unidentified flying objects. This testimony he gave for the purpose of refuting the Condon report that the evidence against the existence of U.F.O.'s is conclusive. At this point in the press conference, Professor Charney sprang to the defense of McDonald, stressing the high repute in which Professor McDonald is held by his colleagues. "Dr. McDonald's opinions on the U.F.O. were unorthodox, but should be regarded as evidence of his honest and fearless pursuit of the truth."

Why did Charney support McDonald? Because of McDonald's effective opposition to another panel on Weather and Climate Modification. In 1964, that panel, on which sat some prominent physical scientists, handed in a report[7] which was highly pessimistic about the possibility of weather modification. McDonald fought this conclusion persistently, forcing the panel to re-examine the issues, including consideration of additional facts which McDonald adduced. The result was the report of 1966, No. 1350[8], which in its conclusions was the opposite of the previous report, No. 1236.

I believe that the erosion of the status of the physical scientists is to be traced to this report No. 1236 of 1964, which apparently was compiled with less than normal thoroughness. The lessons to be learned from these events are:

1. That scientists sitting on government panels must be ready to devote more time to the task, or must refuse to participate.
2. That the community of physicists must assume collective responsibility for the operation of these panels and should certainly seek to be informed about them.
3. That this clearly requires expenditure of more time on public problems by each physicist than he has allocated until now.

Our concern about present-day society's attitude towards the physical sciences is underlain by a desire to turn time back to the Niels Bohr era of 1920 to 1950. That era was governed by the spirit of Niels Bohr, and had its own standards of behaviour and their associated demands on the individual physicist. If we want a return to the spirit of that era we might adopt the guiding principle of asking ourselves how Niels Bohr would have reacted to each issue, and try to react in a like manner. For example: since, in the Bohr era, physicists rarely followed the current practice of devoting a day a week to industrial consulting, it may be necessary henceforth to devote that free day instead to problems of the impact of the physical sciences on society. This will put a particular burden on the surviving disciples of Bohr, some of whom are with us today, who will have to show the way by personal example.

Fortunately, physics has also prospered in less favourable environments than the Bohr era. If we look back into the history of the impact of the physical sciences on society, we find, for example, that Isaac Newton, on April 16, 1676, felt compelled to write a letter[9] to the Secretary of the Royal Society, Oldenburg, warning him about the possible social harm that Newton foresaw in an invention published by Boyle[10] in the Philosophical Transactions of the Royal Society of February 21, 1676.

The title of Boyle's article was *Of Incalescence of Quicksilver with Gold,* and therein Boyle related that he came upon a recipe for an extraordinary mercury which "purged by Sublimations and Distillations" could perform alchemical incalescence, i.e. could react with gold powder to develop heat. In this letter to Oldenburg, Newton raised the prospect of unnamed dangers to mankind if the alchemical truths Boyle was said to have possessed should ever be announced to the uninitiated. The motives of this concern for the public welfare were probably not altruistic. It is suspected that Newton, who himself was heavily engaged in alchemical experiments, was concerned lest Boyle had the jump on him in discovering the Philosopher's Stone by which the alchemist dreamed of transmuting metals.

On the other hand, we know that Newton urged his disciple, David Gregory[11], to destroy a model of an invention made by Gregory's father for improving artillery on the grounds that it would soon become known to the enemy and that it tended to the annihilation rather than the preservation of mankind.

One final remark. My account may have given you the impression that physicists have failed in their duty to society, and that we must conclude that public affairs should be taken out of the hands of the physicists and should be entrusted perhaps to writers or poets, because they are unbiased.

I reject this idea, and shall cite one example to support my view: an incident in the life of the poet Ossip Mandelstam.[12] In 1922, Mandelstam was caught in Wrangel's front in the Crimea. He managed to extricate himself and ran north. By the time he reached Petrograd his clothes were torn. He was advised to file an application with the Union of Soviet Writers for clothes coupons. So

Ossip Mandelstam	Judgment of Maxim Gorki,
wants	Head of the U.S.W.:
1 shirt	
1 pair of trousers	He can manage without them.

I do not want to entrust my fate to poets who may apply overly delicate criteria in deciding which part of a scholar is worth preserving.

---

[1] *Senate Hearings, Civil Supersonic Aircraft Development (SST)*, H. J. Res. 468, Fiscal Year 1971. U.S. Government Printing Office, Washington, D.C., 1971, p. 428.
[2] Ibid., p. 429.
[3] Ibid., p. 434.
[4] Ibid., p. 472.
[5] Ibid., p. 476.
[6] Ibid., p. 250.
[7] *Scientific Problems of Weather Modification*. NAS-NRC Publication 1236, Washington, D.C., October 1964.
[8] *Weather and Climate Modification, Problems and Prospects,* Volumes I and II, NAS-NRC Publication 1350, Washington, D.C., 1966.
[9] Manuel, Frank E., *A Portrait of Isaac Newton,* Harvard University Press, Cambridge, Mass., 1968, p. 178.
[10] Ibid., p. 176.
[11] Ibid., p. 288.
[12] Mandelstam, Ndezhda, *Hope Against Hope,* Atheneum, New York, 1970, p. 117.

Discussion

Prof. Leo Picard, Hebrew University, Jerusalem

I am neither a physicist nor a chemist, nor a biologist but a geologist. I share your interest on the subject of the present and future relation between science and society. When listening this morning to the lectures and discussion, I sensed a certain undertone of pessimism as to the fruitful influence that science might have on society. However, in this afternoon's session there prevailed a far more optimistic mood among the speakers. I am no less optimistic. The reasons may be due to my work in Israel for the last 47 years, during which long period I was forced, *nolens volens,* to bring the application of scientific knowledge into close relationship with society.

The present conversation as to the impact of science on society might as well be reversed. In other words, the influence of society on science or better on the scientists should also be analyzed. An example may be taken from my own experience in Israel.

Whenever a new agricultural settlement had to be established, and at that time large parts of the country were still semi-desert, advice on the proper location for groundwater wells was sought from us geologists. It was this pressure put on us by society that led to the deployment of our utmost scientific and technical efforts and to the discovery of unexpectedly rich aquifers which fundamentally changed the development of our country.

I doubt that we scientists could have achieved these results without the constant drive, financial risk and optimistic approach that stemmed from responsible leaders of our society. Members of the Weizmann

Institute and of the Hebrew University can certainly add similar cases in which the economic and social demands of the population have successfully stimulated a responding scientific contribution from them.

If science is to have a decisive influence on society, then scientists should not stay in their self-contained ivory tower but widen their contacts with people outside their narrow surroundings and extend their intellectual dialogue to the openminded heads of the governing bodies of their country. With this co-operative exchange of views in mind — not only in war time but also in times of peace, of which Prof. Pekeris demonstrated such impressive examples this afternoon — the question as to which influence is more important, that of science on society, or vice versa, remains but an academic one. In this context Brussels, the seat of the European Economic Community, has certainly been well chosen as the place for this conference.

Prof. Sir James Lighthill, Cambridge, England

The question of the SST that Prof. Pekeris mentioned is a very good example of where the environmental effects have been studied in a too specialised way and with far too little looking forward. It has been the tendency in every country where the SST problem has been discussed to look only a short time ahead and to ignore the environmental impact of quite a different kind, that of the shrinking of the globe. The noise aspects have certainly been grossly misrepresented by those whose interest has been to terminate these projects. It is very important at the outset of such projects to take the broadest possible view of all environmental implications, and that is what the French and British aerodynamicists did, ten years ago, when they embarked on the Concorde project.

The economics are, of course, easily confused if one concentrates only on the first few supersonic transports from the production line. But because jet engines can operate so much more efficiently at supersonic speeds, there is the possibility that supersonic transport will be more economic than the subsonic transport and provide transport at three times the speed, at the existing costs per passenger and per mile.

We are conscious in France and Britain that after the first 50 or 60 Concordes have been sold and used by the airlines to make supersonic travel popular, we shall be able to produce developed versions of them that will give the great boon of supersonic transport to the world at a very economical price. One reason why our project has been so far more successful than the American is that we took the right technological decisions, and this is very important. The crucial decision was to go twice the speed of sound and not three times the speed of sound with its many additional hazards.

The point about the altitude, that the American SST would go 21 km inside the ozone layer, as against the Concorde at 17 km below the ozone layer, is not, I think, important, but it is nevertheless relevant to current controversies. What is much more important is the use of materials which operate well at the Concorde speeds, and the general engineering simplicity that was possible at this lower speed, avoiding such complicated features as variable sweep back and so on.

I have been closely in contact with the Russians during the development of the two projects and we were delighted by the way in which the same problem produced in the end a sort of convergent evolution of technological solutions, although initially there was an immense divergence. The conclusion I am trying to come to really is that a degree of restrained optimism is possible in these fields where there is determination from the outset, as Prof. Pekeris said, to follow a goal, with an analysis from the beginning of all the consequences there will be from following the goal by different technological routes.

Prof. Michael J. Higatsberger, University of Vienna

I would like to come back to the main theme of our discussion. This is not the question of whether something is going 17 km high or not; our theme is the impact of Science on Society, and I think the weight of our discussion should be on the word "impact" and on the word "science". The impact of science, in my opinion, has the components we have already discussed, but there is one component which was not even touched upon, namely the economic impact of science.

I think to a great extent, the current questioning whether science is good or not good, whether we should continue or should have a moratorium, is actually about economics. There is apparently not enough money available nowadays to continue the atomic age or the space age, but there are governments ready to sponsor and support these endeavours. It seems that some of the institutions and companies involved in these projects were hiring people, some of whom could not be regarded as scientists. The problem was intensified when at some of the richer institutions, the unions or others on the staff, rather than the scientists, were really the ones that made the decisions.

In trying to come to a conclusion I think we should acknowledge that efficiency in science is an important element of today's discussion.

If we come back to efficiency, I think the question of science being good, or not good for society will disappear very quickly.

Dr. R. Aron Brunetière, President, French Committee of the Weizmann Institute, Paris

I should like to express briefly the ideas of the man in the street. I am not a mathematician, nor am I a physicist or a chemist. I am a doctor. For 30 years doctors have been a sounding board for measuring the impact of science on society. For 30 years we have watched the average length of human life prolonged, to all intents and purposes, by 20 years, largely due to penicillin and to the antibiotics which followed in its wake. And in the course of these years, when I have treated and on occasion cured a patient, I have not asked him to understand the chemistry of penicillin and its biological mechanisms. Nor did I ask him to understand the chemistry of cholesterol and steroids when I prescribed cortisone derivatives.

This morning I have heard eminent speakers evince many complex questions. There was talk of rigorous self-discipline, and of limiting the progress of science. I don't believe we should limit the progress of science, nor do I believe it can be limited. I believe that science by its very nature is bound to progress constantly.

Truth is what it is, its own force and its own virtue. Having said this, the impact of science on society is in actual fact this progress we are witnessing in the domain of health, it is this progress in the comfort and enjoyment of life that science promotes. I did not say in human happiness, for that would be another subject.

Discussion

But there is another aspect, one which does not concern only the scientifically under-developed countries, but also nations as developed as our own. This is the difficulty which scientists find in communicating with non-scientists. It is a difficulty all the greater because scientists belonging to different disciplines often have trouble in understanding one another.

As a result we witness an attitude of mistrust, sometimes of withdrawal, sometimes of blockage, on the part not only of the ordinary man but his leaders as well. For the physicists, even more than for other scientists, this distrust raises problems and explains the worry expressed this afternoon by Prof. Pekeris among others. For physics needs to be understood by society if it is to continue to get the large budgets needed for the type of expensive research it pursues.

There is therefore a problem of communication to be solved. It is the problem of trying to convey to those who don't follow us.

To do this we must find a language which will permit scientists to demonstrate that it is in the general interest to allow science to continue its natural progression.

Prof. Victor F. Weisskopf, M.I.T. Cambridge, Mass.

I would like to say a few words in the name of the man in the street about the SST question. Right away may I say I am not an expert and I have been on none of these committees; I have just read the newspapers and I am living in the second half of the 20th century. The decision of the American Congress was not, I think, so much based on expert opinion, given in a very naïve way as Pekeris has described it to us, but on the feeling of what is technical progress and what does it serve.

The question is: Is faster transportation so important just because it is better than the previous kind of transportation? The man in the street sees that so-called progress has brought him the clogging of the streets, the pollution of the atmosphere in the cities, and he asks himself how should we now deploy our admittedly strong technical capabilities? Shall we really spend so many, many millions to make the trip from America to Europe shorter by a few hours, or should we not rather try to use our knowledge to improve the world we live in— this of course is to a great extent not a scientific problem but a social problem.

I think it is a matter of priorities, and it is very important to ask whether technical progress is *per se* a desirable step. And I think that the doubt about this was the main basis of America's decision regarding the SST. I should like to add that I am by no means a partisan here, I am an observer and interested in the opinion of this group which has a lot of supporters of the SST. Is it really always to the good to pursue all technical possibilities when obviously many of them have brought us to disagreeable positions, to say the least!

Prof. Friedrich Cramer, Göttingen

We cannot talk, I think, at this level of optimism or pessimism because the question of the impact of science on society is a scientific question and should be treated in a scientific way.

Discussion

Prof. B. Eckmann, E.T.H., Zurich

In discussing physical sciences, I feel that not only physics, technology, economics and education should be considered, but also fundamental science. I am a mathematician, a very pure and abstract mathematician, yet all my life I have fought against the separation of pure and applied mathematics. It is quite clear that mathematics, a fundamental science in all aspects, has an influence on society. One aspect is important, namely that this influence is most of the time completely unexpected.

Let me give a few recent examples. I would say that when Boolean algebra was invented, nobody would have thought about its application to computers. Or when category theory was invented in order to put mathematical theories on a solid, real structural basis, nobody would have thought it would have implications for automata theory and for economics. These implications are extremely strong, only they are not so well known even among scientists.

And the third example is "new mathematics"; probably you do not know much about new mathematics, but children know it, and explain it to their parents. When set theory, the foundation of mathemathics, started nobody would have thought that a language was being developed that is of greatest importance to the understanding of science. Yet here in this new mathematical training there is a language which can be understood by everybody, whatever his home background, whatever nation he comes from, in whatever underdeveloped nation he lives, and this will have its effect very soon.

Here then are some examples of the unexpected nature of the influence of fundamental science on society in a very broad sense. Now what conclusions should one draw from this? One very important conclusion I think is that there is in such a field no science policy; it is impossible to programme such a field, or let's say, it is infinitely more difficult than one would think. We could of course try to predict the unpredictable like the metereologists, or say, let's simply hope to do the right thing for the wrong reasons, and then we should simply get together and find the wrong reasons, agree on them, and then everything would be fine.

But, joking aside, we have to be aware of the fact that scientific methods alone, Professor Cramer, are not sufficient for this. Here optimism, trust, belief, are necessary, and certain objectivity alone is part of our scientific attitude, but it is not part of the biosphere. And I would simply repeat what Prof. Weisskopf said this morning "we owe to society to continue in science despite of all the dangers that have been mentioned". The main thing is the attitude and the education.

Dr. B.J.A. Bard, The National Research Development Corporation, London

I only venture to speak because I think the remarks I want to make very much endorse what the last speaker has said. As far as I can see the scientist essentially provides information and it is impossible for him to tell in advance whether or not society is going to use it.

Of course the work may have been carried out with an objective, be mission-oriented as we say today, for making money on the one hand, for helping society on the other, or it may have been carried out simply to increase our fund of knowledge. But in the end society will pick from that pile of knowledge what it thinks of value. In this selecting process there are many, many factors at work, some of them

Discussion

irrational psychological factors, interests and prejudices, quite apart from the normal factors of economics.

This leads me to a second point: the involvement of the scientist in the impact of his subject. Perhaps he becomes involved here because he feels that his work is of value and he simply cannot put it on the pile and wait for it to be picked up. Once he becomes involved in this way, he is no longer a scientist, but someone engaged completely in the mechanism of transfer or, if you like, an entrepreneur on behalf of that information in order to get it through the impact barrier.

What is worrying me a great deal at the present time has to do with what was said this morning about greater asceticism and cutting down growth. If we cut back I do not see how we can avoid increased unemployment in many countries because so many jobs depend on industry. Should industry keep at the same level, it will need fewer and fewer people because, as a result of science, it will make itself more efficient. Hence a minimum growth level is necessary to employ the same number of people.

Therefore we can say we need more industrial research to have more growth to maintain employment, and thereby provide money for the underdeveloped and developing countries. If we do this we can get into all kinds of other pollution and environmental problems. It seems to me that this is the basic problem which has emerged from the impact of science on society, certainly over the last 50 years, and I think this is the kind of problem in which the wisdom that Prof. Sabin referred to this morning is very, very much required. One hopes that it is in this area, of how the impact works and how it can be cushioned, and how it may be sustained and made sensible, that the scientists will be able to make an important contribution.

Prof. J. A. A. Ketelaar, University of Amsterdam

There are, I think, no limits to science, nor to the curiosity of the individual. However, if Galileo had had to ask the Pope for an enormous sum of money to construct the telescope to investigate the idea that the Sun was the centre of the Universe, the Pope certainly would have refused because he undoubtedly would have thought he had more worthy projects to support.

A further question arises, what limit is there to the amount of money and labour we are going to put into different projects? Once you have to construct, let us say, a super-CERN, how far can you go? Not that I feel there are no interesting and new phenomena to be discovered, phenomena of which nobody can foresee the possible applications. The point is that we shall have to make choices, and I think the main difficulty now is how to make these choices.

We have come from an age where the individual professor could more or less do what he wanted for a limited sum of money, but that time has passed. The limitation now is not science, and not curiosity, but the total amount of available money and labour, for all the things we want to do. And it will always be the case that for each project there will be only a very small group relative to the total population that is fundamentally interested. The degree to which they have to convince the others of the worthiness of their project is a matter of education.

Prof. Hendrik B. G. Casimir, Chairman

May I perhaps venture one or two remarks. First of all, one might ask whether in itself the fact that big projects are uneconomic might not in the long run be a good thing in the over-developed countries. If I understand it rightly, the pyramids in Egypt were not only built as monuments, but also to provide work for the farmers during the months their fields were flooded by the Nile. I do not know exactly about the labour situation of cathedrals in the middle ages, but it has been said by others, and I would subscribe to it, that building a big accelerator can in a certain way be compared to building a large cathedral.

Now one might say this has one great advantage; you don't produce an over-abundance of goods that don't really contribute to the dignity and happiness of human beings which is the case in many of our modern industrial activities. Also, the amount of air pollution produced by such accelerators is slight, and the amount of material consumed compared to the world resources is small. Perhaps since we cannot stop technology the only really advanced technology that can be tolerated by society in the long run is this cathedral building. These huge installations require a great deal of work and money, and do not produce anything but pure knowledge and pure science.

There was a remark by Dr. Bard about production growth being necessary to keep the economy going. This is certainly so, but it also means that we are using our natural resources to an ever greater extent. I sometimes imagine that our technology in years to come, if the world keeps developing, will be one of such perfection that machines will need only a very limited amount of power to do what today requires large amounts of power. At the same time perhaps we shall take pride in both using less and less material and in eschewing conspicuous consumption.

If we are willing to pay the price for such an ideal society it could perhaps employ more and more labour without causing an overloading and overtaxing the world's resources. These are perhaps idle speculations and forgive me for abusing the privilege of a Chairman to say a few words.

Professor Pekeris referred to what he called "the Bohr Era", and mentioned there were pupils of Niels Bohr in the audience. Professor Weisskopf is one of these pupils and perhaps he would make a few remarks.

Prof. Victor F. Weisskopf, M.I.T., Cambridge, Mass.

Just let me remind the Chairman that he is an older pupil of Niels Bohr than I; at least I found him in Copenhagen when I arrived. I would like to respond with a general comment on Pekeris' paper. It was a tale of human insufficiency, human frailty. Whenever humans deal with a problem they are apt to do so in an insufficient way, especially when they are pressed to do things in a certain time. Scientists are accustomed to something else.

The strength of science is that it solves those problems that can be solved, leaving aside for the moment those that look insoluble. Later, many of the unsolved problems find a solution. Now, when scientists are forced to take a position on so-called political issues needing immediate decisions, they come into a position for which they are not trained. I once heard a definition of a "responsible position". It is a job requiring you to make decisions about things you don't understand.

Scientists are not accustomed to this, and when they are forced to make such decisions they make all those mistakes that Pekeris described so colourfully. He said that Niels Bohr probably would have acted differently. I agree, for Bohr for me is a father symbol, the greatest

Discussion 99

personality I have ever met. Therefore I am convinced he would have acted differently from the people Pekeris mentioned.

I think that we scientists are in an extremely difficult position and I don't believe that the school of Bohr or the people who worked with Bohr are any better, or any worse, than any other group of good scientists that was fortunate enough to participate in solving the riddle of the atom. Some of them became a little cocky from that experience— certainly not Bohr. Perhaps some people say: because we solved this riddle we can solve all riddles. Now, it's well known that this theory does not work.

I would like to repeat again that I think what Pekeris says is right. However, one should not draw too strong a conclusion from it, namely that scientists should not try to be involved in questions of immediate decisions because they make mistakes. But may I be a little arrogant? I do believe they make fewer mistakes than others.

Prof. Albert B. Sabin, Rehovot

This morning Prof. Eigen suggested that perhaps it would be helpful if we left the sphere of broad generalizations and tried to derive some laws, or guide-lines for action by examination of certain specific problems. Now we scientists, like everyone else, want to do the thing we love, and in our case it is to produce knowledge to solve the problems faced by society. But the actual solution of these problems involves issues that society as a whole must resolve.

This requires planning by both science and society. As M. Spinelli said this morning, when you begin to plan, you have a set of priorities. And when you have a set of priorities, you have values. Prof. Weisskopf touched on this when he said "Is a mere extension of technology in itself a value thing?" Here is where society and science begin to come together. Scientists are forcing society to set up value judgments, and value judging is a field of philosophy. But the philosophers by themselves cannot be asked to do this.

So that means the scientist must come out of isolation; the philosopher must come out of isolation; all of society must come out of isolation. The fellow who holds the money must come out of isolation too, because money is one of the things that make it possible to live or not to live. We must get used to the idea that money is not a dirty word, that it is the very life blood of every social activity. I am willing to see 10 percent of the gross national product put into scientific exploration. But the allocation of the money is not easy. In order to avoid even bigger mistakes than are being made, I think we should develop working bodies that will try to arrive at the best possible decisions.

Let me give you a practical example that is creating somewhat of a crisis among scientists in the U.S.A. The President and Congress under the impact of social pressure have decided to allot a great deal more money to cancer research, and the scientists are worried. What are they worried about? In the first place, there is a requirement that the fundamental research in all fields of natural science having a bearing on the ultimate problem of cancer must be supplemented by specifically designed programmes that will determine whether the course of the research does or does not have relevance to the human problems.

That takes a totally different kind of planning from that of the individual scientist asking a question and providing the answer, and then leaving it to others to put together. So scientists are afraid that most of the money, perhaps at the wrong time, will go in directions

Discussion

that will yield neither results nor ultimately yield the knowledge that is still lacking. I happen to belong to those who do not agree with this particular diagnosis of the problem. I do not agree because in my judgment it is necessary at a certain time to make decisions as to whether the information you have been gathering has a bearing on the main target.

In the United States 350,000 people a year die of cancer, and what a miserable kind of death; and a million people have to be treated. The people are saying: we have been spending hundreds of millions of dollars for years now on cancer research — where are we? And is there anything we can do?

If we did not have science, society could not survive, and science would have to be invented all over again. If we allow the proponents of a moratorium to eliminate science, we shall have to start all over again, so that is not the issue. Science and society depend on each other, but their alliance has become a terribly involved problem. It is important that science does not receive the blame for the failures of society. And we must make it clear that science must be allowed to continue to develop new knowledge in order to help solve the problems of society.

Who makes the priorities? Where can science help? This can be made into a much better world only when political leaders develop plans to utilize science better and in the right places. We talk about utilizing natural resources: for the last 25 years the standard of living in the U.S.A. has depended not on natural resources; it has come from the technological developments initiated by scientific discoveries. New industries created new jobs, but how long can the developed countries go on merely serving their own market? They will have to develop markets all over the world, for unless the hundreds of millions of people in the developing world can be brought to a level where they can buy the products of the industrial nations, the latter will find themselves in a war of competition for limited markets.

The job is not to fight for limited markets, it is to create new markets, and it is going to take technology to create these new markets. Therefore we must work together with society, because it can't do without us. And we, as scientists and humanitarians, are concerned in helping to solve human problems, but we are not in a position to make the decisions of the values or priorities.

Prof. Léon Van Hove, CERN, Geneva

From the many comments I would like to take four and devote a few words to them. They are not taken in chronological order. The first one comes from Prof. Ketelaar, who raised a very important question of how much money and effort has to be devoted to the big science. My reaction to this is that I split the problem in two sub-problems. I believe that the first thing one has to know is how much total money is to go into pure research of all types.

The second question then is what will be the relative priorities of that budget. I am quite convinced that we, the scientists, have been spoiled by 20 years of wealth, and we have not a selective policy, but an additive policy of science. In the old days a little research group was under one professor, with one assistant and a few students, and when the professor was tired of a problem or unable to solve it, *he* made the decision to drop it and move on to another problem that was more on the frontiers of research.

Discussion

Science has now changed its structure considerably, but the process of selective decision in science policy has not been developed. We have been very fortunate in having a steady growth of research budgets of well over 10 percent per annum over the last twenty years, so additive science policy has worked. This will not go on much longer; in fact it is over.

We, the scientists, are again having to learn to be selective. We must stop what is no more than a continuation of rather uninteresting things, and go on to things that are at the frontier, be they big or small. Regarding the very large instruments which some people, (I must say I am not one of them) compare to cathedrals, it is crucial to remember that their construction cost is a minimal fraction of their operation cost during their operating years. The real question is not their size, it is their level of exploitation and their selectivity; we must employ them on really worthy experiments and have the patience to wait for the new idea, rather than experiment before we have it.

Somebody raised the question of efficiency of research. This is, in my opinion, very closely related to what I have just said. The collective nature of research and the huge supporting staff involved often force the scientists in charge to jump to the next problem to avoid leaving their technical staffs idle for six months while the leaders take time to think. In addition, as you all know, the extra-curricular activities of the research scientists involving much travelling has meant that in thinking about tough problems they often postponed, and postponed too long.

The third point concerns a remark by Prof. Eckmann. I am very glad that he intervened to stress the immense impact of mathematics on human affairs, and I am sorry that this was not given a more important place during this meeting. One other point which was also stressed and very rightly so, concerns the impossibility of predicting what discoveries there will be and what will be their consequences. Now, how does that fit into planning? I think the answer here is that we should not be too pessimistic, because science proceeds by asking those questions that are small enough to have hope for an answer.

We do not jump to the final questions straight away, and the small questions allow us to predict to a degree what means, what tools, what facilities will be needed to make a little step forward in research. What that step will be — a result, no result, useful results or bad results — that is another matter where prediction is impossible. But a certain predictability in the operation, the preparation of tools, both human and material, is possible. Nevertheless we should always stress the fundamental impredictability of the results of science and of their consequences.

And finally I hope my comments show that I agree with Prof. Sabin that we should go to concrete possibilities as soon as practicable.

Prof. Hendrik B.G. Casimir, Chairman

This brings us to the end of this session. I believe that both the lectures and the remarks that have been made this afternoon give us considerable food for thought and will give those who have to summarize the whole meeting tomorrow afternoon quite a task. As they will do it, I do not have to try to summarize this session. I have only to thank the two speakers and all those who contributed to the discussion.

Remarks

Dinner on Evening of Monday, June 28, 1971

Chairman:

Sir Siegmund G. Warburg

Address on the Same Occasion

Professor Raymond Aron, Membre de l'Institut

Sir Siegmund G. Warburg is President of the merchant banking firm of S. G. Warburg & Co. Ltd., London. He was born on September 30, 1902 in Tübingen, Germany, and educated at the Reutlingen Gymnasium and Urach Humanistic Society. He is a member of the Board of Governors of the Weizmann Institute of Science, and has assumed the Presidency of the European Committee of the Weizmann Institute.

Remarks

Sir Siegmund G. Warburg, London

Assembled here tonight are a considerable number of distinguished scientists, politicians, writers and industrialists and I have quite an inferiority complex in attending as a banker. From my own experience I look upon bankers not as professional people in the deeper sense of the word but rather as some kind of professional dilettantes or amateurs.

However it is perhaps a modest justification for bankers in a gathering like to-night's that we try to be friendly intermediaries and enthusiastic onlookers and progress chasers. Be that as it may, it gives me great pleasure to welcome, on behalf of the European Committee of the Weizmann Institute, a group of important friends of our Institute.

It is now my duty to read to you a telegram which we have received from Mr. Malfatti, the President of the Commission of the European Communities: "I am truly sorry not to be able to attend the Symposium arranged in Brussels by your Institute which is renowned throughout the scientific world and which I have visited personally. My regret is all the greater because the subject selected for the Symposium *The Impact of Science on Society* is a highly topical one. I am profoundly convinced that our destinies depend largely on our ability to reconcile scientific and technological progress with the safeguarding of human values and in this spirit I wish you and your Symposium every success."

It is not my business to talk about the scientific or organisational aspects or about the spiritual background of the Weizmann Institute.

In my capacity as I see it, as a pure amateur, in the double sense of the word, I should just like to express some strong personal feelings which I have about the Weizmann Institute and which link up with some of the things which were said to-day during the Symposium about the many variations in the meaning of the terms "Impact of Science on Society" and "Impact of Society on Science".

I recollect certain impressions of our late friend Weizmann on this very point. He took with equal seriousness his work as a scientist and that as a statesman. Although he has probably contributed more than any other human being to the creation of the State of Israel, I always had the feeling that to him scientific work was at least as important a part of his activities and of his mental make-up as his political aims. I have heard him say on more than one occasion that, while the political and military organisation of Israel was in his eyes a matter of continuous urgency, Israel's dynamic initiatives in spiritual matters were to him the supreme mission of that small country.

For that man who was in emergencies able to be an extremely practical and a highly pragmatic politician, ultimately the over-riding priorities were spiritual matters in their widest range, from human relations and emotions to the fields of art and science and in these two fields quite particularly art for its own sake and science for its own sake, "l'art pour l'art" and "la science pour la science".

With specific reference to the Weizmann Institute and its endeavours Weizmann often expressed the fear, which I think Professor Weisskopf mentioned, namely that fundamental science might be neglected by too much emphasis on applied science. Weizmann felt that the pure scientist should be simply involved in the search for facts, whilst the application of such findings should not be primarily the scientist's business. Only very reluctantly was he prepared to admit at times that even in science detachment can go too far.

We heard much on these lines this morning. Indeed the fact that utilitarian application of science can go too far is a point which is a special and significant part of Weizmann's important heritage. This heritage is still very much alive and has been an invigorating factor in the Institute's approach to big and small matters alike. It is the reason why I think we are entitled to speak about the Weizmann Institute not only as a research and teaching Institute in the usual meaning of the term, but as an Academia in the best sense of that word as it was used in Greek thinking and Greek philosophy.

I know that it is the ambition and the passion of the workers at the Weizmann Institute, from President and Chancellor downwards, to keep continuously at the most intense heat possible the *feu sacré* of the academic spirit and to make thereby a unique contribution to the community of Israel.

Professor Raymond Aron, Membre de l'Institut and of the Collège de France, has had a distinguished academic and literary career. Born in Paris on March 14, 1905, he studied at the University of Paris and became a Doctor of Literature in 1938. During the war he was editor of *La France Libre* in London and after the war on the staff of *Combat* and *Figaro*. Subsequently he held academic appointments at Tübingen, Harvard, Cornell and the University of Paris. He is an honorary Ph.D. of Harvard, Basle, Brussels, Columbia and Southampton Universities.

Evening Address

Prof. Raymond Aron, Paris

One of the organisers of this gathering, having convinced me that the presence of a sociologist with a philosophical background would not seem out of place amongst real scientists, sent me, to encourage me, to instruct me, to provide me with a model, or perhaps to remind me of due modesty, the text of an address comparable with the one it is my pleasure to deliver this evening.

This address was given on the October 21, 1947, by Dr. Weizmann, first president of the Republic of Israel. It had as its title, "Modern Man — Slave or Sovereign?"

It opened with the following words: "I am embarrassed by a double difficulty in addressing this distinguished Assembly on the subject assigned to me. I am a scientist, not a philosopher. Moreover, I must express my views in a very few minutes, and though I were to speak with the tongue of men and of angels, I could hardly hope to contribute anything of significance to the subject."

I should say I am neither a philosopher nor a scientist, and although I have at my disposal more time than Dr. Weizmann, I feel, compared to him, such an inferiority that I must confess my embarrassment even more than he did.

What can I say on the relations between your scientific disciplines and society that you do not know as well as I do, and which has not been said and repeated by all the newspapers and on all wavelengths? Today's techniques have multiplied the means of transmitting messages more quickly than the number of messages of significance. Repetition and redundancy make a noise that besieges and fatigues the spirit — different from that made by motorist or transistor, but perhaps just as pernicious.

Since the leaders of the Weizmann Institute wished me to speak this evening in circumstances where solemnity and entertainment are mingled, when each one of you at the close of a day of study wants to forget his narrowly professional interests, I was obliged to look for the reasons for a choice which does me honour but whose motives remain to me a mystery.

There is no shortage of other speakers who could with advantage have filled my place. Rightly or wrongly, I attributed a particular intention to those who invited me. Since the explosion of the first atomic bombs over Hiroshima and Nagasaki, there is scarcely a scientist who has not reflected on the relationship between science and politics.

But the reflections of an Israeli physicist or a sociologist of Jewish origin and French nationality, looking beyond these simple contradictions, crystallises the extreme situations in which the essential is laid bare, that is to say, the often tragic quality of our situation.

"Master and Owner of Nature", the Cartesian formula keeps its symbolic value. It is of little import here whether the scientist maintains the distinction between truth, aimed at and attained in itself and for itself, and the uses of knowledge, or whether, on the contrary, he admits that by its very nature the knowledge due to modern science incorporates its own application.

Nuclear physics, for a quarter of a century, and perhaps to-morrow biology, have demonstrated that apparently pure theories, removed from all practice, make possible the manipulation of elements, and that this manipulation has not, as such, any other end than that fixed for it by its users: destruction or construction, arms of apocalyptic war or electricity generating stations.

Not by accident but by necessity, man achieves over matter a mastery that is still only partial, yet at the price of a sometimes excessive growth of his power for evil. Perhaps, indeed, the negative, destructive use of technology normally costs less than its positive use. It costs less to raze a town than it does to rebuild the ruins, less to intoxicate masses with propaganda than to educate the individual; it costs less to prevent men dying by introducing modern methods of hygiene eliminating microbes and epidemics, than to give them the resources necessary to lead a decent life.

To these simplifications certain scholars reply that they are seeking truth for its own sake and that only the search for truth *Aristotle dixit* — is good in itself. These people consider themselves morally authorised to pursue in their laboratories or their offices the solitary undefined quest. Others, to make up for it, suddenly awakened to the realisation of their responsibilities, have crossed the threshold of the market place: politics, they have thought, is too serious a matter to be left to the politicians. The *Bulletin of Atomic Scientists,* now called *Science and Public Affairs,* and the Pugwash Conferences, bear witness to this participation of citizens, educated within the rigorous discipline of science, in the debates of the market place.

Is it a participation that is sterile or fruitful? I shall not allow myself to discuss the matter. I shall limit myself to a remark which in itself, it seems to me, rises above the dispute: In the dialogue on the most effective method of avoiding nuclear catastrophe, the words of the physicist do not contrast markedly with those of the political theorist or the politician.

The most famous physicists sometimes oppose each other passionately. They all knew in 1949 how a thermo-nuclear bomb might function; they did not know for certain that they would succeed in making it; they know today with certainty how it works. The President of the United States probably does not know, but this ignorance does not mean that one must choose the President of the United States from amongst physicists. They, as physicists, do not possess the diplomatic art of playing with a menace they do not wish to put into operation.

Nor does the political theorist, for, unlike the physicist, he does not possess either the knowledge of his objective which deserves the description scientific, nor, still less, the technique of its manipulation, founded on proven learning.

In short, when the scientist begins to interest himself in the use statesmen make of the arms which he has provided, he hesitates between two contradictory feelings: One is a revolt against the dealings of the politicians in the shadow of the nuclear apocalypse, a mistrust of knowledge which is pretentiously called "political science", and which, for certain, hardly resembles the physics of particles, and the resolution to influence the decision of the powers that be.

As an alternative, the scientist resigns himself to the austere destiny of the researcher in the frozen universe where truth is discovered, always good in itself, and where instruments are forged flexible and at the behest of their users.

About 30 years ago, scientists took it upon themselves to instruct the statesmen; the letter from Einstein to Roosevelt remains in everyone's memory and retains its symbolic value. At that time a kind of unanimity united the scientific community against the Hitlerian enemy.

Not one of them could imagine without horror the atomic weapon in the hands of him who already appeared as a demagogue, if not a monster. At the worst moment of the cold war the same unanimity was never re-established.

Whatever may be the individual's judgment on Stalin and on the Soviet Union at the time of the personality cult, the Soviet Union, quite differently from the Third Reich, had never raised to the level of dogma the reprehensible principles which scientists, in the exercise of the ethics of their profession, must reject with scorn.

Already doubt began to gnaw at the conscience of the scientist in the service of national defence, therefore of the cold war, therefore of a state certainly democratic, but who knows, perhaps dominated by the forces of wealth or monopolistic capitalism.

The war in Vietnam transformed these doubts into a revolt at least for a number of the scientific community in the United States. According to custom, two questions posed themselves, linked, but inseparable in law. Did not certain means dishonour the army which used them? Was the war being waged by the Republic a legitimate war?

Between the judgment of the means and the judgment of the end, no link was necessary. The bombing of Dresden in 1945, when full of refugees, appears to us, now that the fury of combat and vengeance has abated, as useless cruelty.

The movement, whose initiative was taken most often by the youth of the United States, to break off relations between the Department of Defence and the Universities, bears witness to a fundamental paradox: certain scientists can only do their research properly when working for the Army, the Navy or the Air Force. When they condemn the politics of their nation they feel torn both by the political condition of the man and even the citizen, and they feel it more keenly than anyone.

The refusal to participate will perhaps give them a moral peace, nothing more. The Minister of Defence, if university laboratories close their doors to him, will provide his own laboratories. If students no longer join the reserve of officers, the Army will consist of nothing but professionals, and the American Republic will in its turn follow the road taken by so many republics which have become great and corrupt by imperial tactics.

Understand me clearly: I do not allow myself to make judgment, I strive to understand. The scientist, as much as, or more so than any other citizen, has the right or if you like the duty, to evaluate the conduct of the State which draws benefit from his work. When he finds himself in radical opposition with the regime or the external politics of his State, he has the right and perhaps even the duty to demonstrate, by abstaining, by the withdrawal from the future or even by active protest of his inner convictions.

These diverse attitudes are not without their effect, in certain circumstances. But in a democratic State, governed by men elected according to the laws, the scientific community will rarely speak with one voice in its refusal to serve the State.

Even if the majority of the members of this community disapprove of a certain aspect of the external politics of the State, certain of them will hesitate to press home their refusal to serve, for fear of disarming their own Republic which must endure whilst wars in Vietnam—and here I use it as a symbol—pass away.

The scientific community does not therefore act as such; certain of its members act, and legitimately, as citizens do, by word and by motion. They do not and cannot resolve the dilemma; as a citizen, the scientist shares the common destiny, his knowledge does not measure the influence he exerts on the decisions of those in power. Other pressure groups jostle in the corridors of the Assembly.

And so the scientific community depends henceforth on the State in the same way as the State depends on it. Certain research, called "Big

Science", claims financial resources which only resources of the State can provide.

The scientific community has also a second reason for being divided. At times split by what judgment should be passed on the State it is condemned willy-nilly to serve, it is divided also inevitably since a choice must be made between possible research in terms of the means offered. The allocation of sparse wealth, if the market does not look after it, and in this case the recourse to the market is excluded, rests with the power of the State.

It is the State, in the last analysis, which for a large part establishes the budget of research and development and it is the men of science, or those endowed with power by the scientific community, who try to convince those responsible and try to obtain a distribution which ties in with the general interest or the interest of different branches of learning.

The difficulties of a comparison between intrinsic value or the value for the nation of different pieces of research, the plethora of considerations radically heterogeneous condemn the relationships between the scientific community and the leaders of the State to equivocations, more typical of political life than the truly scientific life. The days of the Ivory Tower and of Innocence are past.

An effective science, a science indispensable to military power, to the wealth of nations—for such is, we know, in certain of its important parts, modern science—becomes an affair of State. Now at last here are two activities, each obeying its own law, obeying a specific logic, each as demanding as the other in respect of those who dedicate themselves to them, dangerously near the one to the other.

Gaston Bachelard wrote in a striking formula: "To overcome men with men, is sweet success in which the thirst for power of the politician takes pleasure. But to overcome man with things, that is the enormous success in which no longer the thirst for power triumphs, but the shining thirst for reason, *der Wille zur Vernunft*". This thirst for reason, and the thirst for power, may they never be confused!

Chaim Weizmann, I believe, would have liked Bachelard's Formula. Scientist and politician both at once, he would not have denied the necessary distinction, he would probably have added that, in the last analysis, the hope of human perfectibility, in which he saw the very substance of the Jewish message, forbade accepting this opposition as final.

Why not dream of human friendship in a common thirst for reason? Certainly — but perhaps the word reason has not the same sense according to whether it is a question of agreement among researchers or a compromise between opponents.

Two adjectives mark the difference: rational and reasonable. It is rational to bow before the evidence of fact or demonstration, it is reasonable to take account of circumstances and resign oneself to the inevitable.

According to a saying of Herodotus which I like, no man is sufficiently devoid of reason to prefer war to peace, because in time of war it is the fathers who bury the sons and not, as in time of peace, the sons who bury the fathers. And yet how many, many are, by Herodotus' definition, devoid of reason!

The Institute which organised this gathering remains, you have just been told, a centre of scientific research in the grand tradition of the Universities and Academies of Europe and of the United States. The attendance here in Brussels of so many eminent personalities therefore presents first and foremost a significance of an intellectual and cultural order. The members of the Institute belong to the trans-national community of science.

The smaller a country is, the more its sons must look beyond its frontiers to maintain communications with their colleagues beyond land and sea. The Israelis would not be able to fall back on themselves nor even to pen themselves up in the Near East without betraying their vocation and placing it in peril. They belong to the civilisation which is called, for want of a better term, Western. A number of its manifestations — science and technology — belong already to the common treasure of the whole of mankind.

But the same Institute which bears the name of a scientist and of a man of politics, conceived by him as a centre of unbiased research, how can this Institute escape the constraint of the tragic years if one remembers the spiritual finality of the new or renascent State?

The search for truth, in certain cases an effective truth, namely pure science, but also, science in the service of national defence, provide the answer. How could it be otherwise in the world as it is?

Israeli scientists find themselves in a situation comparable with that of the Western scientists during the last war rather than that of American scientists today. Agreeing with their state on essentials, they do not refuse to work for it, trusting that it will make a morally acceptable use of the instruments they provide. By this means, whatever we may say, this meeting of scientists runs the risk of taking on in certain people's eyes a political aspect or character.

Those of our students who repeat that nothing escapes the clutches of politics, who accuse their masters of ulterior motives, hypocrisy or naïveté, remind us at least of a trite proposition when a conflict between classes or people reaches a certain extreme violence, namely the freedom of work and of human activities—an essential condition of any authentic culture—risks being shattered by the wave of passions.

In certain countries at war, more than a half-century ago, even the music of Wagner became suspect, and extremist patriots even came to suspect Kant or Hegel of being at the root of Pan-Germanism or the crimes of Prussianism.

The conflict between Israel and the Arab countries has created and continues to create in France and elsewhere a jealous climate between those who align themselves with one camp or the other. In this atmosphere every action, every gesture, every word is immediately blotted by political significance. Why then, if you will allow me, should I not seek with complete frankness the ultimate reason for all our presence here this evening, understanding that this probably varies from person to person. Of course, I am only speaking here for myself; I am thinking aloud before you.

This meeting of scientists to discuss scientific problems, together, constitutes first and foremost, just as every such gathering, an act of faith in the universal vocation of science, the affirmation of a community which may be divided by frontiers, but which in spirit recognises no frontier.

For a quarter of a century, because science carries with it the springs of power, the States, which have remained sovereign, have time after time forbidden or delayed the spread of knowledge which they have defined as military secrets, and not without good reasons. And I mean from the legitimate point of view as those in charge of national defence are struggling to withhold from a potential enemy knowledge which is translateable into instruments of destruction.

That this sometimes frantic care over concealment has in the last resort not prevented what those responsible feared is not sufficient reason to bring to an end the debate between the peaceful community of scientists and the warlike society of sovereign powers. The one refuses on principle to keep secret any truth once discovered, the other does not wish on principle to declassify knowledge for state reasons. It is a dispute that will only end the day frontiers disappear. Until that day comes scientists must set aside political conflicts, whether great or small, with an eye to a free interchange of ideas.

There is no need to have an opinion on Israel's frontiers or on the future destiny of Sharm-esh-Sheikh to take part in this exchange. It is enough to think — and where is the scientist who does not — that the scientific community excludes no one. Certain people exclude themselves from it by their own decision or under pressure from the powers. Whoever speaks the language of reason, Israeli or Arab, has a place in it and will be made welcome by it as an equal.

There we have, I think, the first meaning, the most obvious one if you like, of our presence here this evening. Scientists must safeguard, as far as they can, the trans-national city whose freemen they are. This city knows the quarrels of the different schools of thought, even of vanity, but refuses the conflicts of nations as defined by their frontiers. According to Hobbes these nations permanently keep their powder dry and their guns loaded because they mutually suspect each other of the darkest plots and, alas, not always without reason.

Even when centres of power no longer recognise each other, the scientific city remains open to all citizens of all nationalities. Yes, a gathering of this kind takes on a political sense, but this political sense consists primarily and above all in refusing to let hostilities between states, however passionate they may be, break up the scientific community and interrupt its communications.

I am not sure that I shouldn't stop there; and yet if I did I would feel I was sacrificing frankness to discretion. As a Frenchman of Jewish origin, to use a familiar expression, I received and accepted an invitation which had far more significance than an invitation to any other congress. The Weizmann Institute keeps a place for the humanities, for the social sciences. It wants to keep alive the tradition of the European universities, of the best among them.

The so-called natural sciences, or in English simply the sciences, represent today the great human venture, the reason for working, the pride of modern civilisation. Without them there would be neither cultures worthy of their name, nor nations capable of survival; they neither form the total of our knowledge, nor the only inspiration of wisdom.

The spirit of revolt which labours in Western society against certain consequences of the Industrial Revolution, against the uses made by our society of power and riches, reminds us, if it were necessary, of the millennial lesson which the Holy Books offer us in symbolic language. By losing the innocence of non-knowledge, humanity threw in its own face a challenge which has not yet been accepted, which it will never finish accepting.

This argument would probably not have been enough to determine both my invitation and my reply if the Weizmann Institute had not borne the name of the first President of the Republic of Israel which embodies the spiritual vocation of the Jewish State; and if I myself did not belong to the diaspora — I am a Frenchman who wants to be French and who is mysteriously linked to Israel by bonds whose exact nature I myself cannot determine.

Among the Jews of the diaspora let us begin by saying that there is not and cannot be a common attitude towards Israel. Young Jews, leftists, were militant in the Palestinian movements. The special issue of *Les Temps Modernes* devoted to the Israeli-Arab conflict carried an article written by a Middle Eastern expert of Jewish origin, who gave the most serious and well summarised report of the Arab arguments.

Religious Jews of the American Council of Judaism, non-religious Jews of the left, or extreme left parties, do not feel towards Israel either fellow-feeling or even solidarity. We have no right to condemn them in the name of principles which they do not accept. Each of us is born fair or dark, tall or short, French or German, into a Jewish family or a Christian family. But if each one, whether he wishes or not, is forced to accept his size or the colour of his hair, it is by no means the same thing for the nation or Judaism. Henceforth, the Jew of the diaspora chooses his own nationality and he chooses it in relation to his birthplace and at the same time in relation to Israel.

The choice of non-Israeli Jews arouses a certain difficulty, depending on national demands and their totality, whether ethnology and religion on the one hand and national allegiance on the other seem normal or not.

There is a first snare to avoid: that of the bad conscience and hyper-compensation. The Israelis, citizens of their democratic country, criticise their government. Why should the Jews of the diaspora think themselves obliged to support Israeli rulers in all circumstances, why should they fear being accused of betrayal?

I had no sympathy in 1956 for the Franco-British expedition to Suez; I said so and I don't regret it. If as a Frenchman I criticised more than once the French Government, I cannot put aside my right to criticise the Government of Israel, or I would be betraying myself and the political values for which Israel stands.

Attachment to Israel only implies a sort of unconditionality: the right of Israel to existence and recognition. Anything less raises political thoughts, that is to say in the best case reasonable and not rational, probable and not certain, preferable but not good.

At the most pathetic moment of the war in Algeria in 1957 I wrote a little book in favour of Algerian independence which created a scandal. I was quite well aware of the consequences of independence for Jews in Algeria who were French citizens. In certain extreme situations every decision affects certain groups cruelly.

To satisfy the Algerian desire for independence was a just cause but that decision brought in its wake injustices for families whose fathers, grandfathers and sometimes great-grandfathers had lived in that land from which they were now hounded.

In 1956 at the time of the Suez expedition, in 1957 on the subject of Algeria, certain of my writings were translated into Arabic by Egyptian newspapers and writers. I feel neither shame nor regret on that account. I endeavoured to see clearly and by good fortune I have never been aware of a radical incompatibility between justice (or what passes for such, or near enough, in political affairs) the interests of France, and those of Israel. But of course there is nothing to guarantee that compatibility.

Have I the right to draw a lesson from my experience in the course of these latter years during which, writing in the daily press as a sideline from my teaching activities, I have refused the security of the Ivory Tower? It would be vain to believe so. At the very most, some ancient truths will regain their freshness for us.

Citizenship I believe responds to a kind of moral need. Human existence includes a political dimension. But citizenship risks shutting us in, it limits our horizons. How can we serve our motherland without abandoning any of our freedom of judgment? How can we serve it without betraying it or ourselves? The Israelis can no more do without the exchange of thought with the Jews of the diaspora than they can exist without the exchange with Israel.

The Israelis are at war with the Arab countries; the Jews can and must in good faith, maintain relations with their friends in Arab countries in every way, wherever these Arab friends will allow. How can all of us, meeting here at the behest of a scientific institute, not aspire to enlarge our circle and dream of the blessed moment when beyond deaths, sufferings and humiliations, Arab and Jew, as one individual to another, will regain contact, will look each other in the face and find a common language—and what language more ready for this dialogue than that of science?

For centuries, Jews prevented from bearing arms by the condition imposed upon them, went about as strangers against an heroic background. It was a strange prejudice which did not recognise the heroism of patience, humility and hope, long suspected, even if it was only by themselves, of cowardice. How could they not have throbbed with emotion, with pride, at the military success of the Israelis.

The world had forgotten the reputation of Jewish fighters during the Roman Empire. The Six Day War gave rise to the myth of the "Prussians of the Middle East"; but whoever has seen the disarray of the Israeli soldiers' uniforms and the loose relationship between their officers and men, saw at once the absurdity of this myth.

Perhaps even I said to myself the day after the Six Day War, "Now the image of the Jew, unable to fight, dedicated to trade and money, will disappear for ever." If I had that thought, and I probably did, I put it down to human weakness.

History obliges Israel to live dangerously, to survive thanks to the force of its arms. It dictates neither its vocation nor its message.

Dr. Weizmann, in the address I quoted at the beginning, said the following words which I should like to recall: "We are confronted by the strange paradox that a character creates more lasting values than achievements. What a man *is,* means more in the long run than what a man *does*. The same is true of nations."

And Justice Frankfurter of the Supreme Court of the United States defined in these words Dr. Weizmann's ideal: "A Jewish State for Dr. Weizmann means the creation of new moral and cultural values."

How can we doubt that these two men would speak with the same voice today. And with them all the friends of Israel and especially all the Jews of the diaspora in the knowledge that a people, inseparable from a religion, will only remain faithful to its dreams if it can define its spirit, or, to speak like Dr. Weizmann, because it will *be*, more than because it will *do*.

Malraux during the Second War exclaimed: "May victory remain with those who waged war without loving it." Israel has waged war without loving war. In the words of the Israeli soldiers, collected after the Six Day War, there was no trace of hatred against the enemy, no ecstasy of triumph.

What mockery if the Jews of the diaspora drew vain glory from the successes of war by the Israelis, if they did not benefit from the security behind the lines in which they live to escape their base passions, if they did not rise above the legitimate passions of those risking their lives in the front line.

Every scientific institute these days serves national defence and that bearing the name of Weizmann is no exception to the rule. The Institute is not for all that a means, it remains an end — it is not an instrument but a symbol — the symbol of that community of reason and truth, open to all Christians or Muslims, Arabs or Israelis, Jews or freethinkers.

It invites us to cultivate in ourselves the spirit of peace, that is to say to respect the other man and his otherness, to recognise in him the equality which we were for long denied, so that one day, beyond the forgotten battles, beyond the honouring of the dead, reconciliation will become possible — the only lasting victory, the only victory worthy of those who founded the Weizmann Institute and the State of Israel.

Morning Session

Tuesday, June 29, 1971

Chairman:

Prof. Wolfgang Gentner, Heidelberg

Professor Wolfgang Gentner has been the Director of the Max-Planck-Institute for Nuclear Physics in Heidelberg since 1958. He was born on July 23, 1906, in Frankfurt a.M. and studied at the Universities of Erlangen and Frankfurt from 1925 to 1930, receiving his Ph.D. from Frankfurt in the latter year. From then until 1939 he held fellowships at the Institut du Radium de la Sorbonne, Laboratoire Curie, and the Radiation Laboratory, University of California and was a Scientific Assistant at the Kaiser-Wilhelm Institute for Medical Research at Heidelberg. Subsequently he was Professor of Physics at the University of Heidelberg and the University of Freiburg and from 1955 to 1959 was Director of Research of CERN in Geneva.

Morning Session

Chairman: Prof. Wolfgang Gentner, Max-Planck-Institute for Nuclear Physics, Heidelberg

I have the honour as a physicist to introduce to you two biologists, Prof. Maaløe and Prof. Samuel.

In the discussion yesterday, the relation between science and education was mentioned, and I think Victor Weisskopf was also talking about teaching the man in the street. In this connection I would like to mention the name of Amos de Shalit, for he would certainly have been extremely pleased to be here and discuss this problem with us. Amos de Shalit was a brilliant theoretical physicist in the Weizmann Institute whom I met in the 1950's at CERN. We had long discussions about the exchange of young people between Israel and Germany, and in this way we came to the co-operation in certain ways between German and Israeli institutes. He was also very interested in teaching problems. In his later years (he died two years ago) he built up the science-teaching centre in Rehovot which is now called "The Amos de Shalit Science-Teaching Centre". Also the main high school in Rehovot is called the "Amos de Shalit School". I only mention his name because he was a man who thought it extremely important not only to do research, but at the same time to teach people what research is.

Yesterday we had long discussions about many problems, particularly the moratorium question and the priority question. Perhaps today in the discussion after these two lectures we could also talk a little about teaching problems, if we have the time.

Now I would like to introduce to you Prof. Maaløe, one of the founders of EMBO, who is now the Head of the Institute of Microbiology at the University of Copenhagen. He is a member of many academies, and is going to talk about his own research field.

Professor Ole Maaløe is Head of the Institute of Microbiology at the University of Copenhagen. He was born in Copenhagen on August 15, 1914, and received his MD in 1939 and Ph.D. in 1946. From 1948 to 1958 Professor Maaløe was Director of the Department of Biological Standards at the State Serum Institute of Denmark.

Can Ideas from Molecular Biology be applied to Economic and Social Systems?

Prof. Ole Maaløe, Institute of Microbiology, Copenhagen

My contribution to this symposium was prepared with the idea, which I think must be common to all the speakers, that each of us would begin somehow within his own field of research, and then, ideally, converge on the main theme. I should probably have preferred as a general title: "The Interaction between Science and Society"; I think this would have allowed for all the different ways you can look at the question. But, of course, "The Impact of Science on Society" has more force and direction.

Now, I am going to follow the line that I have sketched from my own research into the theme of the symposium very literally. To do so, first of all I should tell you that the main interest in our Institute for years has been to learn something about the rules and regulations governing the growth of bacterial cells, these being the smallest, and probably the simplest, organisms that we can study today.

I will begin by relating to you an incident that occurred some years ago and which may give you an idea of what I am driving at. I asked a prominent economist the following question: Suppose you were asked to design a flow diagram for a big production plant about which you knew nothing. Am I right, I asked him, in assuming that you would go to the plant and ask to be shown the most complex and costly component in that factory, and that you would begin your work by stipulating that this unit, which might be a mechanical device of some kind, a computer or a transport system, must never be idle and must always operate at its maximum efficiency? Now, my friend the economist looked at me with what I took to be surprise at the naïveté of my question and then he said: But of course! The answer was not unexpected to me, and in turn I could tell him that this is exactly the principle on which bacteria operate when they grow.

In these extremely small organisms hundreds of different parts of the whole are constantly being produced in proper proportions so that, between successive divisions of the cell, the number of each of the many components is roughly doubled. However, all these very well regulated activities depend on the presence of a relatively large and quite complex unit which is called the ribosome. And you will guess, of course, that in my analogy to the factory, the ribosome is precisely that complex and costly piece of equipment which must be operating efficiently at all times.

This unit is in fact a site in the cell on which the hundreds of different enzymes are assembled, and the rate at which the cell is able to grow therefore depends utterly on the number of ribosomes present, and the efficiency with which they work. And the reason I said a moment ago that the bacterium operates like a well organised factory is precisely that, in the course of evolution, the bacterial cell has "learned", in a given growth condition, never to produce more ribosome than it can put to use with very high efficiency. Thus, the principle which seemed so obvious to my economist friend seems to have evolved without the help of the human brain in the very distant past. And I will venture to guess that this may indeed be one reason why it seems so obvious to us.

Now, having stated what the behaviour of the system is like, I want to consider some features of the evolution of this system, but I am not going as far back as Dr. Cramer went yesterday, i.e. to the non-living soup, within which selection was made for something that could reproduce. I am satisfied to go back to the most primitive bacterial cell I can think of.

Of course, we are now stepping on pretty soft ground, because we are talking about something that occurred in a dim past and about events which have left no visible trace in the world we live in today. Still, I believe that some very general, and perhaps useful inferences can be made from an analysis of this kind. In the evolution of life on earth, culminating in something like this meeting perhaps, that which we might call a "modern bacterium" (i.e. one equipped with all the features that we can study in present day cells), represents a highly complex system, exhibiting at least some of the traits characteristic of all the higher forms of life, including man.

Specifically what I am thinking of is the fact that bacteria have evolved a large number of quite refined control mechanisms. The mental exercise I want you to join in begins with an assumption for which there is now reasonably good evidence: namely, that the ribosomes I have mentioned came into being, in the form we know them now, a very long way back in evolution.

I will assume furthermore that most of the control mechanisms which govern the production of all the individual components in the cell, such as the hundreds of different enzymes that catalyze individual acts of production in this factory, have evolved later. At this point, many of you will have realised that I am referring to the so-called "induction" and "repression" mechanisms that were first analysed in great detail by Jacob and Monod. The purpose of all these individual control loops that exist in the smallest of cells is to ensure that the cell, at all times, produces enough, and not too much, of any particular enzyme. In other words, the several hundreds of induction-repression mechanisms in the bacterial cell act in concert to heighten and optimize the efficiency of the total system.

Let us now consider the evolution of a system containing some, but not all, of these many different control systems. All we need to know for the present purpose is that each control mechanism requires that, through evolution, a particular, new protein be introduced into the assembly of the whole cell, a new protein the function of which is to turn on or off the production of a particular enzyme, i.e. of a particular species of all those that are being produced in the factory.

In a mechanical system this would correspond roughly to the regulation of a flow by the proper use of a valve. What I want to argue is that each of these control proteins must itself have evolved, and through mutation and selection it must have acquired that particular sensitivity which makes it a useful control element in the cell. In other words, it must enable the cell to economize on the production of the enzyme which it controls, but it must not over-react, because by doing so it would prevent the cell from producing the requisite amount of a necessary enzyme.

Thus I envisage a very long period of evolution during which one specific control mechanism after the other was introduced into the whole system, and it seems very likely to me that some particular feature of the system as a whole has been governing the selection processes through which the individual control mechanisms, the little control-loops in the big network, have approached the sensitivity needed to serve the overall system in the most efficient way.

It is quite difficult to guess what the specific feature might be on the basis of which optimization has occurred. However, when you deal with bacteria in contrast to society, you can do experiments and with luck you can get a good hint as to what this main feature of the system might be. As everybody knows, antibiotics and drugs of all kinds are important in medicine; what is less well-known to non-biologists is that each antibiotic affects some particular function in the cell, and that antibiotics therefore can be used as refined biochemical tools.

We have recently worked with a new antibiotic, called fusidic acid. You will recall that I started out by saying that we have seen that the ribosome system is under very effective control, and that ribosomes are made in the requisite number so that all of them can function with high efficiency. Fusidic acid was interesting to us because it specifically reduces this efficiency. If you study the effects of this drug, you find that the poisoned cells react in a "sensible" direction — i.e. to counteract the reduced efficiency, they make more ribosomes; however, they overreact grossly. The control systems, which operate with great accuracy when the ribosomes are allowed to perform in their normal way, over-react in the presence of fusidic acid in a way that does not serve any obvious purpose. In other words, the cells do not know how to handle this new situation efficiently.

Now I want to recall to you the fact that the ribosomes seem to have existed in their present form for a very long time, and I want to suggest that the efficiency of the ribosome has been a determining factor during evolution. What I mean is that the highly evolved bacterium we know today has gone through an evolutionary process in which the individual control elements of this complex system have adjusted to one another, and that this adjustment has always been made with reference to the ribosome and its efficiency in protein synthesis.

If this is true, it is not surprising that fusidic acid, which alters this very basic parameter in the cell, creates a situation in which the total control system, although it reacts in the right direction, does not operate efficiently. In trying to understand such a complex system I think we can use this sort of observation as guidance and look for some feature of paramount importance in the system around which everything else has been organised. If such a feature can be identified we may eventually be able to understand the system, and perhaps even to modify it.

Now I have to make an enormous jump, and if I said before that in thinking of the evolution of the bacterial cell we were treading on soft ground, I'm now definitely plunging into deep water. What I want to do is to use the exercise we just went through to consider interactions between science and society, and of course I want to try to guess as best as I can what feature in today's society might be of such overriding importance that it should be kept in the forefront of our mind at all times, even when we try to cope with an isolated and seemingly major problem, say, the war in Vietnam. Of course, we now depart from the bacterial system except as a model.

First let me say that I am not trying to draw parallels to human genetics, or the actual control mechanisms of a biochemical nature, such as hormonal controls. I shall move all the way into the field which some people have called "epi-genetics", that is to say the kind of biology which is not directly governed by DNA. What I am thinking of is that we learn from others, that we have a history which is communicated from one generation to another, and the fact that within this structure of learning and communication a very large number of controls on the behaviour of society operate.

It seems to me not too far-fetched to argue that, as we legislate, or adopt rules of conduct, we are in fact introducing into the medium of epi-genetics one control feature after another which together define the actions and reactions of society. Of course, this pattern of controls can be and is being modified slowly; to modify it drastically and quickly is perhaps the greatest challenge and the greatest difficulty that man could conceivably come up against.

The magnitude of the problem is perhaps most evident when we consider that the present state of society is the result of a slow evolution during which all the "controls" I have talked about have been deeply embedded not only in our literature and history, but in the very language used to transfer these controls from one person to another.

As a medium our language has become loaded with meanings and associations from which it is extremely hard to escape. Here I want to limit our discussion to the last three or four centuries of Western civilization, because I think there is one particular feature which has been free to rule the behaviour of man and society to a very large extent. And that is the feeling of accomplishment, of the enormous value that man has attached to what in a very broad sense may be called "conquest" — conquest of ideas, of territory, and of power. This may sound trivial or self-evident to some, but that is no reason for taking it lightly.

Yesterday, Aharon Katchalsky mentioned that the world we are living in was, largely speaking, an open system till not so long ago, and that it is increasingly becoming a closed system. It is this fact which constitutes the great challenge. In an open system it is possible to live with an exaggerated thirst for conquests of all sorts. In crises, of which there have been many, the solution lay in expansion, that is, people went elsewhere to be able to do what they wanted to do, or what their thirst for conquest told them they wanted to do.

This, we all know, is not as easy to do as it used to be, and soon it will be quite impossible. Therefore I think we must be bold enough to realize that if we go from an open to a closed system what is required of us is a very basic change in our value system. In terms of the particular stimulus, which I have called "conquest", it must be de-emphasized since it can no longer be exercised with the freedom or recklessness to which we have become accustomed.

Finally, I would like to comment on the matter of teaching, namely, that it would be especially useful in the present situation if man in general knew more about science and how science operates. I fully agree that this would be good, but I cannot help thinking of a point made in a recent book by Ivan Ilitch. He says very bluntly that if a Latin American is privileged by being sent to the United States to go through a college education, some 300–400 times as much public money will be invested in his career compared to that of his friends from a family of middle income.

There could be no objection to this if the world at large could be imagined to progress smoothly to the higher level of education. What I want to point out is that from the point of view of the third world, the fact that the U.S.A. or the Western societies in Europe are willing to invest that amount of money in order to bring a man through to a certain level of knowledge, indicates that they see an advantage in doing so; and the feeling is that gifted people from "outside" are "bought" or captured by the *status quo* system.

There is a great deal of suspicion that the higher forms of education obtainable in the U.S.A. or in Western Europe are a subtle way to bring people over to the side of the *status quo* system which obviously, from the point of view of the third world, has to be changed. This is a real and a valid concern which we simply need to know how to overcome.

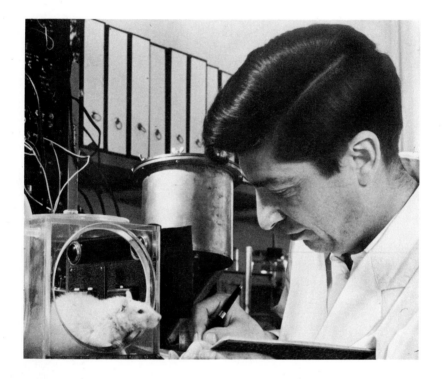

Professor David Samuel holds the position of Sherman Professor of Physical Chemistry at the Weizmann Institute. He is also Dean of the Chemistry Faculty and Chairman of the Board of Studies in Chemistry of the Feinberg Graduate School. Professor Samuel was born in Jerusalem on July 8, 1922. He studied at Oxford and received his Ph.D. from the Hebrew University of Jerusalem in 1954. In 1957–58 he was a postdoctoral fellow at Harvard and in 1965–66 a visiting fellow at the University of California at Berkeley. His research interests include mechanisms of reactions in solutions and at surfaces; the methodology of the teaching of science and the chemistry of the brain and behaviour.

Science and the Control of Man's Mind

Prof. David Samuel,
The Weizmann Institute of Science, Rehovot

We have heard a great deal at this symposium about the multiple problems of science, about money and moratoria, about priorities and education of the public, things to which our distinguished Chairman Prof. Gentner has alluded. I would like to draw attention to a problem which was hinted at by John Kendrew in his opening remarks, that is, that much of scientific endeavour has become either dull or rather dangerous. It is true that some areas of research are no longer as intellectually stimulating or as able to capture the imagination as in the past, that many scientists have indeed moved on to new and more challenging fields and that, as you have heard, a number of physical chemists have switched to biology or solid state physics. Not long ago, about five years ago, I myself moved from a field which is considered by many to be dull to a field, brain research, which is generally considered to be potentially dangerous.

I shall not talk today about the philosophic aspects of this research, such as the relationship of brain and mind, but rather outline some of the problems, touch upon some of their implications and try perhaps to answer the questions raised by Albert Sabin yesterday: What are you, the scientists, trying to do *to* us? And what are you trying to do *for* us?

The field has many other names: it has also come to be called "psychobiology" or the biology of the psyche, although of course it contains large elements of chemistry and not a little physics. It is a field of such obvious significance in terms of the future of humanity that it has been accompanied by a chorus of warnings against the impending control of individual will, not necessarily for the good. This feeling of impending doom has been taken up by the press and other media, presumably because doom sells better than destiny, and the gloomy view of life is more newsworthy than an optimistic approach.

Unfortunately, this has coincided with the general tendency in public opinion to be anti-science, and as Aharon Katchalsky has hinted yesterday, even to be anti-intellectual. As a result we are now faced with the suggestion that there should be a moratorium on certain lines of research. Part of this attitude is, I believe, a reaction to the pressures of the material world and a general disappointment in the achievements of science and it is, I think, not unrelated to the increasing use of drugs in the Western world.

The "drug culture", in fact, has emphasised the dramatic effect of chemicals on the brain, and the various ways in which chemistry can open up the "doors of perception", and close the "doors of reality". By now, every one knows the perils of the uncontrolled use of drugs, particularly addictive drugs, and the inevitable wave of crime and extortion that follows in their wake.

But we must not forget that through the use of drugs a way has been found, at last, to alleviate pain, to reduce anxiety, to end insomnia, to control violence in the mentally disturbed and to compensate, in part, for the loss of concentration in the ageing. At the end of the first decade of the science of "psychopharmacology", which is the marriage of pharmacological chemistry and behavioural science, remarkable prospects lie ahead of us. These include not only the sedatives and stimulants that I have mentioned, but also drugs capable of suppressing aggression, reducing fear and perhaps even capable of enhancing specific functions of the brain, such as the ability to learn and to remember. But these effects are a far cry from changing the more complex functions of the brain.

It appears to be extremely unlikely that drugs alone will ever suffice to make people change their minds or vote in contradiction to their convictions. I am not unaware of the phenomenon of brain washing, which is undoubtedly abetted by medication, but it also requires an inter-personal relationship, i.e. that is *two* people—the interaction of an interrogator and a victim, and in order to succeed brain washing cannot be accomplished by a capsule or a syringe alone.

For many years there was little contact between those working in the hard sciences, particularly chemistry, and those dealing with the problems of the mind and the brain. This is a curious failure, because Freud himself considered for a time the possible biochemical aspects of psychoses and other abnormal mental states. Unfortunately, the psychologists who later dealt with human problems considered only man's interaction with his surroundings including his family and fellow men, and ignored the molecular aspects of the brain.

Chemists on the other hand had little use for the early gropings of psychology toward theories of mental activity, of emotions, drives and of learning, which appeared to be based on rather intangible concepts. This dichotomy between scientists and psychologists has only recently been bridged, and today a new interdisciplinary science is being forged. But the existence of this lack of interest for so long a time added to the confusion in the field.

To compound the confusion, the general public tended to feel that mind-affecting drugs were in a way a food like vitamin pills, and thus not basically harmful, a placidity only recently rudely shattered by the well-publicised effects of the hallucinogens. On the other hand, the idea of implanted electrodes in the brain was immediately frightening and caused a far sharper reaction from the start as Aharon Katchalsky mentioned yesterday. It is interesting to consider why there is such a deep-felt reaction to electrical stimulation of the brain, or ESB as it is known, which, in fact, merely changes the chemistry in a well-defined area of the brain by the use of an electric current.

Nonetheless, the sight of a monkey or a cat with electrodes protruding from its skull still causes an instinctive anxiety. Perhaps this is due to years of stories and films about robots and mad scientists, or perhaps it is due to some profound fear of tampering with the brain, of extracting something through the skull by means of wires. It is as if there was an inherent feeling of extreme vulnerability, of protectiveness towards one's innermost thoughts and feelings, which is violated by the idea of an electrode.

Whatever its origin, the possibility of the control of the brain by means of electrical stimulation has become a serious public concern. The resistance to ESB has been heightened by the knowledge that one can elicit "sham" rage in animals by means of electrodes, that Olds and Milner managed to stimulate animals to eat without being hungry, to drink without being thirsty, and that experimental animals and even some men can now elicit some undefinable pleasurable response by electrical self-stimulation of the medial forebrain bundle.

Finally, there is also the dramatic demonstration by José Delgado, at Yale University, of evoking fear, aggression, and even changing social dominance in animals by remotely controlled electrical stimulation of the brain. But Delgado himself, in his recent book *The Physical Control of the Mind,* has made very clear the limits of this technique. He writes: "ESB is non-specific... and this functional monotony rules out the possibility that an investigator could direct a subject toward a target and induce him, like a robot, to perform any complex task."

The flexion of a limb, the control of tremor, the remembrance of things past, even laughter or the desire to excel, can be radio-controlled. But, he says, "Language, culture, personal identity and free will are probably beyond the theoretical or practical potential of ESB... We cannot modify political ideology, past history or national loyalties by electrically tickling some secret areas of the brain". It appears therefore that the fundamental modification of man's mind by chemical or electrochemical means is not possible.

Turning to another area, there has, indeed, been increasing concern recently, following the successful transplantation of hearts, about the possibility of transplanting human brains, which Professor Cramer described in science-fiction terms yesterday. There have already been experiments of this type in simple animals, particularly those in which nerve regulation is extensive and tissue rejection minimal.

In fact, at the Weizmann Institute some experiments of this sort have been performed in amphibia and parthenogenic fish, but this is quite different from transplanting mammalian brains, with their enormous complexity, immunochemical protection, and not much evidence of nerve regeneration. I very much doubt whether such experiments would ever succeed even with parts of a human brain.

Another area of public anxiety, triggered off by extensive publicity, is the success of B.F. Skinner's operant-conditioning. As you all know, by means of a technically ingenious control system and the use of a specially designed box, the behaviour of animals can be gradually altered in accordance with a predetermined schedule of reward or punishment. What emphasises the eeriness of this technique is that rats, pigeons or monkeys can be conditioned to press levers or peck at buttons to get a pellet of food or to avoid an electric shock from the floor, something which none of these animals would do in its natural state.

This of course has led people to speculate on the possibility of conditioning men and women to change their behaviour by similar manipulation of reward and punishment. There have indeed been attempts to get people to stop smoking by the judicious application of electric shock and similar, if more ambitious, attempts to rehabilitate sexual deviants by conditioning. It has even been found possible to get volunteers to change their heart rates, the alpha-rhythms of their brain, or even the pattern of paradoxical sleep, all basic functions which, up to now, were considered to be beyond human control.

But it should be emphasized that conditioning experiments require special equipment, unique conditions, and can only be done with the initial co-operation of human subjects. The very complexity, diversity, and independence of the human mind are, I think, sufficient safeguards against mass conditioning by dictators or madmen.

This leads us to a related fear, which I shall only mention briefly, of control and manipulation of the individual by mass media, due to the power of words and pictures, and the familiar effectiveness of repetition and exaggeration. Perry London, a professor of psychiatry and psychology at the University of Southern California, in a book called *Behavior Control,* says: "Ever since Eve and the serpent began their first fateful talk, the transmission of selected information has remained the most important means by which people have manipulated each other." He lists education, prayer, rhetoric, propaganda, demagoguery, romantic seduction and advertising as typical efforts to this end, all forerunners of the technology which he calls "Control by information". But I feel that this subject, which is not yet a science, is beyond the scope both of this talk and of this symposium.

Perhaps the most significant advance of all, due to basic research on the brain, is the beginning of a rational approach to the alleviation or cure of mental retardation, of many diseases of the brain and the possible conquest of pain. It has been found that many of these diseases are due to defective genes, missing enzymes and blocked metabolic pathways. Now that these processes have begun to be understood, ways are being found to prevent damage to the brain by controlling the diet or by using specific drugs.

Even the alleviation of pain is at last being tackled in a scientific way, and new approaches to this problem have been suggested, ranging from electro-analgesia (electrical treatment of the spinal cord) proposed by Ron Melzak at McGill, to the scientific use of hypnosis, now being studied by Ernest Hilgard at Stanford, and even, as reported from mainland China, the clinical use of the ancient art of acupuncture, in which metal needles are able to block pain in conscious patients.

In the treatment of human defects an exciting possibility is the potential use of direct electrical connection of specific nerve cells to external sensors for sound and for light. These are devices which can convert sensory input into electrical impulses, and may herald the possibility of enabling the blind to see and the deaf to hear—albeit without eyes or ears.

The last frontier in psychobiology will, of course, be the full understanding of the workings of the brain itself; an organ which weighs less than a kilo and a half in an adult, and yet requires a quarter of all the oxygen that is supplied by the lungs and the heart. The energy output of the brain is estimated to be less than that of a dim 25 watt bulb and yet it is the controlling centre of nearly all of the functions of the body, instinctive or learned.

Unlike many other organs, most of the billions of neural cells operate independently, firing and recovering spontaneously, sorting, storing and retrieving information constantly. It has been estimated that there are 200 million telephones throughout the world today, but there are more than a million times more synaptic connections in a single human brain! Considering the increasing chaos of many telephone systems, one gets an idea of the sort of problems that lie ahead of those concerned with brain research, and why progress is so arduous and slow.

In order to understand what changes occur in the living brain, increasingly subtle chemical and electrochemical sensing devices must be used. One can already measure electrical impedance, oxygen tension and even radioactivity in small areas of the brain. By attaching these probes on-line to fast computers it is possible at last to correlate the electrical and chemical changes in the central nervous system with mental activity. By this means, the role and mode of operation of each neuron or group of cells may be elucidated and may suggest a more valid model of the action of the brain for which we are all anxiously searching.

This will lead to enormous benefits in many fields, including education, ranging from new methods for the rapid absorption of large quantities of information, which is now a rate limiting factor, down to specific problems, such as the age at which children are most receptive to learning. The many possibilities of expanding the mind, of memory, of decision-making, and possibly even heightened imagination are no longer as theoretical as they were some years ago before the direct interaction of men and machine become possible.

These are just some of the prospects that lie ahead. There are, of course, some dangers too: from irresponsible experimentation on individual human subjects to attempts to control whole populations by selective information, but they are no more and no less than the dangers confronting other disciplines, such as genetics or nuclear physics.

I have tried to outline very briefly some of the problems and issues involved in brain research. Such research cannot and should not be halted; I most emphatically do not believe in moratoria, I do not even think they are possible. In addition to the obvious benefits of the research that I have outlined, there is the supreme intellectual challenge of understanding the brain itself, of the processing, recording and retrieval of information, the problems of insight of creativity and of dreams.

As I have tried to point out, this research is not without hazard, both in the terms of the individual and in terms of society. No one can predict what new concepts, or devices, will be discovered in the coming years, which will radically alter our approach to the brain, but we can already discern some of the main areas of progress. I feel that it is the duty of scientists not only to determine for themselves what is possible and what is not, but also to make this knowledge available to the public.

I would therefore suggest that the extent and possibilities of research in psychopharmacology, electrical stimulation of the brain, operant conditioning and brain-computer interactions, be studied and evaluated by international and expert teams. This is no easy thing to do, and it is not likely that the opinions will be unanimous, but at least we shall have before us the best evaluation possible of "the state of the art" today, and of the likely trends in the not-too-distant future.

Based on these feasibility studies, I would suggest that a code be established laying down guide-lines for the future, a code to guide scientists working on the brain. It should deal with the ethics of the use of human volunteers for experiments including laboratory investigations of social and psychological pressures, which can be as damaging to the brain as drugs, with the permissible extent of trials of chemicals that affect the personality; and with the moral aspect of transplanting parts of the brain and of reviving patients with severe brain damage.

This code, not unlike the Hippocratic Oath and its variants which guide the medical profession in the preservation of individual men's lives, should help guide us, as the secrets of the brain are steadily unravelled, to preserve the sanctity, the integrity and the individuality of men's minds.

Discussion

Prof. G. Scholem, President of the Israel Academy of Sciences and Humanities, Jerusalem

It is a great privilege to be asked to make some comments from the point of view of an innocent bystander in scientific matters — perhaps not so innocent, but largely innocent of any scientific knowledge. We have heard several highly thought-provoking and interesting papers, especially the two papers by Aharon Katchalsky and David Samuel. Listening to them I have asked myself some questions.

The last paper, by Prof. Samuel, is in a way very comforting and tends to set our mind at rest about some of the more terrifying aspects of brain research. These have been not only on the mind of the general public but came up yesterday in Katchalsky's talk regarding the more sinister possibilities inherent in scientific research and the problems of its control. I should like to make four points which seem to me important from a humanistic perspective:

(1) I would raise a note of warning in connection with the question of values. Katchalsky talked at some length on this point. He spoke about the problems facing society from the application of science and its impact on society, and he said that if there was no way to do research in all directions, somebody would have to fix priorities.

This question of priorities has come up several times in our discussions. If I understand Katchalsky correctly, he said that priorities immediately bring up the question of values. I want to voice a certain reservation. I do not think that priorities and values, on an ethical level, can be easily equated.

Priorities are not determined on the strength of values defined at a philosophical or general level; they are dependent on the specific structure of the society which makes the decisions about them and therefore they are utterly different in different societies. Ethical considerations often play only a minor part in decisions on priorities and we cannot evade this basic fact.

A totalitarian society will establish priorities very different from those that will be made by an open society, and the same goes for a primitive or an industrial society. Surely all these will assign different priorities to problems about which they have to decide.

The question of values remains a philosophical question which cannot be avoided and has to be faced by those who think about science. There is of course a problem by which the bystander and observer of scientific developments is frequently struck when it comes to the scientists *vis-a-vis* the universe of philosophical concepts; namely that the scientists often try to pull themselves up from the swamps by their own bootstraps, a feat which is difficult to achieve.

The problem of values is not one which scientists, in my modest opinion, can solve. Even in the most progressive stages of detailed research, the problem of values will have to be solved by means transcending pure scientific research. I put this to you as the personal view of a non-positivist humanist. We cannot avoid value judgments, but we can form valid ones on purely neutral grounds offered by the scientist. There are no reasons for value judgments inherent in science and people try very hard to escape this uncomfortable dilemma which is not always faced by scientists.

(2) Our friend Katchalsky said yesterday that the globe has become a closed system. We thought it was open, it was infinite but now there is a definite, calculable limit to it. This may be true to a large extent, but I doubt whether it is true in principle. The Earth is after all part of a wider Universe.

The question is how science will proceed to change the amount of available energy and available material (which means energy put to work). This may be, to a larger or lesser degree, subject to changes in some future time by methods which could draw energy from the cosmos, a possibility that is not mere speculation. We should therefore be careful not to exaggerate the sceptical view implied by Katchalsky's words on this point, which to me seems a point of some consequence.

(3) The question of interplay between society and the sciences can, of course, not be solved in a discussion of some hours or two days. The troublesome thing is that these relations exist on two different levels. There are developments brought about by inner processes within

science itself, but there are also dialectical relations between science and society which stem from the dialectics of society, its demands and its inner processes. These dialectical relations, in a necessary process, are putting very strong demands both on society and science. They work both ways.

As long as science has existed such mutual impact has been noticeable. The science of the Greeks had the most tremendous impact on society up to our day. But this impact has not been purely a scientific one. The impact of science on society was always dependent on a mixture of pure empirical research and speculative elements which had their origin not only in science itself.

As far as I can see, no examples are available to show that any scientific system has lacked that fundamental conceptual framework within which it works and which was not only calculated to assist in the analysis and summing-up of the scientists' own findings, but had to do with philosophical and even theological assumptions concerning what they were talking about.

I come back to my first point: you cannot separate these two elements and arrive at a system where science, pure and fact-finding, determines its own conceptual framework. But there is more to this. Some of the speakers made the ironic remark that science was once thought to be an esoteric underground of a few elite people, who alone knew what it was all about. Beyond them there was the *misera plebs contribuens,* the rabble, who had to take it as it was.

Now we are hearing there is a great and wonderful explosion of scientists, even possibly dangerous because — if I understood Katchalsky correctly — in thirty or fifty years the number of scientists needed will exceed the number of people available, for instance in Israel. In contradistinction to this I maintain that the number of scientists who are aware of the implications and fundamentals of what they are doing has not grown very much in its relation to the population.

The number of the esoteric côterie which in the last resort decides on science is as small today as it ever was, and its reasoning is as obscure to the *misera plebs contribuens,* including the main body of the so-called scientists, as it ever was. My feeling is that there is a great difference between scientists and the technicians of science who nowadays are often called scientists and who do not know the first thing about the fundamentals of what they are doing.

I am speaking out of the sad experience of long years with people who came out of the science department of my own university, and I am sure that this holds also for other institutes of higher learning. Here we touch on the problem of teaching. Learning the technique of science is not identical with being clear in one's mind about the process by which scientific concepts are built, by which theories and basic ideas are brought about. I concede that this is a very debatable point.

But I think it should be said by at least one onlooker that, taking into consideration this matter of statistics and the misuse of the word "scientist", the ratio of the real scientists has not changed. I have often wondered at how little people who had studied science, chemistry, physics or biology, knew about fundamental questions which should have been foremost in their minds.

This is sure to be connected with a certain contempt in which concepts transcending the pure sphere of fact finding research, in short philosophical concepts, are held by many people. This contempt is a fact of life, and it is full of grim forebodings for the future of science. The underlying questions facing science education today are: what are the uses of science? What are the meanings of scientific systems and scientific theories?

(4) Allow me a short comment on what Professor Samuel said in his paper on brain research and the control of the human mind. He stated that, for the moment, we can do very little in larger units to affect the human mind by our own interference. This is a comforting thought, but it does not put me at rest. Of course, what we are actually doing in robots, computers and Golems and in brain research on all levels concerns only an infinitesimal part of what is actually going on in the human brain.

Professor Samuel has given a very good image by comparing the human mind with the telephone system of the whole world. But I am afraid that the capacity for harm is greater than he admits. I think this is shown by the fact that even the modest beginnings of biology and genetics which we witness in our time, already reveal possible malevolent consequences of manipulation. I am not inclined to share Professor Samuel's optimism. If some little things can already do so much, how much more is there to be feared from the real dangers for which no inherent remedy can be found within the framework of science itself.

Here again we are faced with the question of ethical problems which cannot be avoided. If at the present stage the possibility of tampering with the human genes is not so remote as is the tampering with the human mind, there is good reason to look with apprehension at the consequences of future results of brain research. The sinister picture painted by Katchalsky should set us thinking.

Prof. Chaim L. Pekeris, Rehovot

My question is addressed to Prof. Samuel. It deals with an event recorded in the book *I Was a Prodigy,* an autobiography of Norbert

Wiener (I see most people are smiling — they have read it). You may recall that after Wiener came back to Boston from Göttingen he fell into a depression. So he went to a psychologist, who told him to lie on a couch and talk — and he talked for several days. Says Wiener: "I tried my darnedest to explain to the psychologist the nature of my personality, until I noticed that that dumb fellow did not understand a word I was telling him, so I walked out on him." My question is, nowadays with the rapprochement between psychologists and physical scientists, what would be the outcome of such an interview?

Laurent Rebeaud, Gazette de Lausanne

I am Laurent Rebeaud of the *Gazette de Lausanne,* a Swiss newspaper. I am, therefore, like some of yesterday's speakers, what has been called euphemistically, a "man in the street". That is to say, someone who is absolutely a non-specialist in your subjects, who has an individual's point of view in every case as compared to yours.

I should like to go back over a distinction made just now between the scientific researcher and the people who are given the job of applying or exploiting the discoveries made by these researchers, in other words the technicians. But not only the technicians, but also the soldiers, the politicians and even the housewives. It seems to me that the scientific researcher is in certain respects in the same boat as the poet is in another sphere. You know that we try, and this is very much the case here, planting on the shoulders of researchers (I speak of pure researchers) all sorts of moral and philosophic considerations.

Now you know that moralists and philosophers are people who, by their very nature, hesitate, retrace their steps, set themselves problems which they never resolve and perpetually contradict themselves. If you like to put it this way, they never arrive at a real result. You probably know what André Gide said about poets, and I ask myself if, in the same way that poets are people a little apart — mad, if you like — scientific researchers ought not by the same token to be considered mad, in other words considered simply as men outside the ordinary norms of society, outside the same moral and philosophic impositions.

And could we not apply the words André Gide used about literature when he said, as you know, "Literature is not made of high minded sentiments". I wonder if researchers should not benefit from the same immunity as the poets and I would ask this assembly — "Do you believe that good scientific research can be accomplished with high-minded sentiments?"

Prof. Aharon Katzir-Katchalsky, Rehovot

I would like to make a few comments both on Professor David Samuel's presentation and on the remarks of Prof. Scholem.

First of all there is some validity in the general feeling that there are dangers inherent in brain research. Some of you may remember a book by Sargeant, *The Battle for the Mind,* which appeared just after the Second World War, in which the techniques of brain-washing, as practised by the totalitarian countries, were described. Some of the techniques were based on direct physiological methods. These early techniques, primitive in the light of our knowledge today, were nonetheless very powerful, as we all know.

On the other hand the development of more and more selective methods, employing chemicals which act on a specific group of brain cells, can produce reactions which go out of control. Samuel was absolutely right in stressing that this research brings the problem of values into very strong focus. With many branches of science one can play around but here there is no room for playing. While there is no question of stopping research in this field, the immediate and far reaching human consequences of such investigations must always be foremost in our minds.

Now, I fully agree with Professor Scholem that the value system cannot be derived from science alone. Science is an integral part of the symbolic system of man, and the developments of science have to grow in full interaction with other branches of human activity. That is why I liked the remark made yesterday by Weisskopf that one should regard science as an integral part of the cultural pattern of human society. Value judgment is an integral part of the same type of development.

Professor Scholem made the point that fundamental science does not grow as quickly as "so-called science". An American statistician of science, Professor Derek Price finds that if the number of scientists is n, the number of scientists producing fundamental results is $\sqrt{n}$; hence the ratio of $\sqrt{n}$ to n decreases with n. From this point of view Scholem is right. But the absolute number of fundamental consequences grows with the growth of science, although the ratio is really diluted.

Professor Scholem made the true point that the development of science in society is not predictable. Indeed, the real difference between biological and social processes on the one hand and physical processes on the other seems to lie in the amount of predictability. Physics, as was pointed out yesterday by Van Hove, has the ability to predict the consequences of actions, an ability not shared by the biological sciences. No biologist would dare to prophesy evolution, or what is going to be the shape of animals and plants in the future.

Discussion

One other point. Even if tomorrow enormous sources of energy become available, as Professor Scholem implies, the finality of the globe, and the interaction between one country and another, still remain within a framework of a closed system. That is why the future planning of human affairs requires a global approach, a system analysis in which local and national planning are not sufficient. From this point of view the impact of science on society means a global approach.

Prof. Ole Maaløe, Copenhagen

First let me try to answer Weisskopf's question. My contribution had two aims: I began with something as primitive as a bacterial cell, because we know quite a bit about it, and I made the bold guess that a major parameter in that cell system has remained constant during evolution, and that all the special control mechanisms have been optimized with reference to their constant parameter.

Then I asked myself: Can this very primitive level of biology teach us something that might be applicable to social systems? I wanted to argue by analogy that even in a social system we should perhaps, before we try to solve individual problems, try to identify a ruling principle which should be kept in mind, at all times, while trying to solve piecemeal the dangers that we see cropping up around us.

Take the Green Revolution as an example. It seemed the obvious answer to a grave problem of malnutrition and hunger. We now know, having tried to put this into practice, that it has very grave side-effects. In fact, when we put the new so-called high yield strains of corn, wheat and rice into use they can only be used effectively by those who can invest in fertilizers, storage capacity and transportation. This often implies that the distance between the wealthy and the poor farmer is increased, because the former can invest while the latter cannot. So, the gulf between the "haves" and the "have-nots" is further deepened.

In this case our first concern should probably have been to consider seriously whether the enormous value which, by historical tradition, we attach to the efficiency-principle may not lead us astray. In fact, I don't think we can hope to analyze, even to a modest degree, what the effects on society as a whole will be of solving piecemeal the dangers and problems confronting us unless we are willing to question the validity of even such a dominating principle as "efficiency".

This comment bears on Professor Aharon Katchalsky's words too, because he talked about systems analysis applied globally, which is a solution in principle only, since, at the moment we simply cannot write the equations. I think it might help if we could discern in the whole pattern of society and its behaviour a general principle of such importance that we would stop trying to solve "minor problems" such as hunger, without due concern for the social system as a whole.

Prof. David Samuel, Rehovot

I shall be very brief. As regards Professor Scholem's worries about optimism and pessimism, I can only quote to him what Abba Eban, Israel's Foreign Minister, said, when he was asked during one of the recurrent political crises in Israel, whether he was an optimist or a pessimist. He replied that he was neither of these but a realist instead. This is what I feel in this case, and what I have tried to emphasize, since I don't think there is a case for either optimism or pessimism.

Discussion

As regards Professor Pekeris' question what would happen if Norbert Wiener went to a psychoanalyst today, I feel that the rapprochement between science and psychology is not of sufficient duration, so I don't think there would be any difference in the result. But I think in 30 years' time there will certainly be a difference.

Finally, to the gentleman who spoke about the immunity of scientists. If the result of the work of scientists were as harmless as poetry, then I think his immunity would be valid. But I think scientists, unlike poets, have a moral responsibility for what they produce.

Michael J. Higatsberger

Professor Michael J. Higatsberger (right) is the Head of the Department of Physics at the University of Vienna. He was born on June 8, 1924 in Unterbergen, near Krems, in Lower Austria, and studied at the University of Vienna from which he received his Ph. D. *summa cum laude* in 1949.

Professor Higatsberger has been a member of the faculties of the University of Minnesota and Graz and was a Consultant at the Army Research Laboratory, Corps of Engineers, Fort Belvoir, Virginia, U.S.A. He is interested in reactor physics and is a member of the boards of the OECD Dragon and Halden Reactor Projects and was Managing Director of the Austrian Study Group of Atomic Energy.

Afternoon Session

Tuesday, June 29, 1971

Chairman:

Prof. Michael J. Higatsberger, University of Vienna

Jean-Jacques Salomon

Professor Jean-Jacques Salomon is the present Head of the Science Policy Division, Directorate for Scientific Affairs, OECD. He was born in Metz on November 17, 1929, and graduated in biology and anthropology from the Sorbonne. Subsequently he taught philosophy in Paris and was science editor of a magazine before joining OECD as Secretary of the Ministerial Meetings on Science. In 1968–69 he was Visiting Lecturer at the Massachussetts Institute of Technology. He has published many papers on international scientific cooperation and science policy and is author of the book *Science et Politique*.

Science and Scientists' Responsibilities in Today's Society

Prof. Jean-Jacques Salomon, OECD, Paris

There has always been an impact of science on society, and there has always been an impact of society on science, even in the days when science as we know it, did not exist. Then it was philosophy, the science of sciences, which exercised an influence on society and society influenced it; knowledge has always been bound up with the social system in which it develops and which it in turn helps to form.

As I listened yesterday to the papers and the resulting comments, I asked myself, what is new in all this? Has nothing happened to give the relation between knowledge and men a new colouring, if not a new meaning? One might have said, at least before Raymond Aron spoke last evening, that science is outside history, eternally protected in its ivory tower, innocent and pure as a new-born babe.

In the corridors yesterday afternoon, an American scientist, for whom I have great admiration and who shall therefore be nameless, said to me that in the United States academic discussions on the benefits and virtues of science, on science for science's sake, would today be inconceivable. On the other hand, here in Europe, with no war in Vietnam, and no pollution crisis, we find that young scientists are questioning themselves and that they are far from convinced that science is for the best in the best of all possible worlds.

No doubt in the time of the Greeks, before science had proved itself, it was possible to be sceptical. Today, when science has proved itself far beyond the dreams of Bacon, Descartes or Newton, all scepticism is ruled out. But we find that among the general public, among the young, and even among scientists, science breeds and nourishes nihilism.

Science has so amply fulfilled its promises of application that a suspicion hangs, not over what it *is,* but over what it *can do*. Not only has science become a political affair, but many people look upon it as the ally of bad politics. Even some scientists think it is in the service of evil and no longer the irrepressible instrument of progress and happiness.

In the conclusion to his book on the history of relations between science and government in the United States, A. Hunter Duprée wrote in 1957: "The mighty edifice of government science dominated the scene in the middle of the twentieth century as a Gothic cathedral dominated a thirteenth century landscape. The work of many hands over many years, it universally inspired admiration, wonder and fear."[1] It is in the United States that science has found itself endowed with the most impressive cathedrals, has enjoyed the biggest throngs of the faithful and has acquired the most effective power of persuasion; but all industrialised countries, since the Second World War, have raised temples with similar ambitions.

Their monuments were not necessarily "the biggest in the world", but they were none the less vaster than anything previously built, bigger in the size of the buildings, the number of scientists, the magnitude and variety of equipment, the amount of resources invested, and noted for the spectacular character of the results they achieved in such a short lapse of time. Thanks to science, everything seemed possible. True, the cloud of Hiroshima raised briefly some misgivings. Oppenheimer said to a visitor in 1956, "We have done the devil's work. Now we have come back to our real job, which is to devote ourselves exclusively to research."[2]

And yet, ten years after this period of unprecedented growth, the cathedral of science is showing cracks everywhere, as though the very foundations on which it was erected were threatened. Enchantment has been followed by disenchantment. Accused of being associated with war and the deterioration of the natural environment and social structure, science finds itself under fire on all sides.

It is attacked from the right by those who denounce it as the costly pastime of mandarins who care nothing for economic return or industrial development; from the left by those who charge it with being the instrument of military and industrial domination, an instrument which pays no attention to the real needs of society and which is all the more pernicious in that it has helped to promote and satisfy imaginary needs.

Our civilisation has developed on the assumption that the progress of knowledge is good in itself because knowledge liberates, and, by its very essence, contributes to the good of mankind. But the triumph of rationality seems to stand reason on its head; reason becomes the support of the irrational. The conquest of the atom plunges us into the balance of terror; the escalation of military power, far from increasing security, multiplies new threats with planetary implications; the discoveries of molecular biology open up the possibilities of manipulating heredity.

Man walks on the Moon, but comes back to find that the problems of Earth are still outstanding and that disequilibrium is growing; two-thirds of mankind has to cope with underdevelopment with less than five per cent of the world population of scientists and engineers. Yet while these less favoured societies dream of factory chimneys, saturated motorways, and ribbon-developed towns, the advanced societies, confused, talk of a moratorium on discoveries and a return to the idyllic nature of Rousseau's man before the progress of science and the arts.

What has happened to make suspect the institution that most strikingly embodies the rationality of the West, and to turn it into a political affair? Traditionally, politics has been a problem for science, but science has been no problem for politics. What has led to the difficulties, tensions and conflicts which characterize the relationship between science and society and which present a challenge not only of politics to science but of science to politics? It is as absurd to hold science responsible for what it has created as it is to expect it alone to provide the solution to the problems with which we are faced, both today and in the future.

Of course, a moratorium on knowledge is absurd; science has a history and one does not stop history. The example of the SST, of which much was spoken yesterday, shows that society can question technology in the name of other priorities. There is a great difference between the moratorium proposed by Newton to Boyle which was mentioned yesterday and the desire to direct research towards objectives which conform to the wishes of the community. The difference is that research has become, whether scientists desire it or not, a political affair.

What must properly be called the ideology of science forbids thinking of it in the equivocal terms of politics; the discourse of science is neutral, a neutrality guaranteed by the objectivity of the method which presupposes rigour, attention to facts, respect for proof. Nothing seems more alien to science than that "tale full of sound and fury" which is politics. The fight against authority, religious, economic or political, is no less a part of the history of science than are its theories and discoveries.

At the dawn of modern science, when he launched the axiom "Knowledge is power", Francis Bacon omitted to mention the difficulties which such an association might involve. The omission may have been deliberate, for being the apostle of a new concept of science, he first had to establish its legitimacy and not to lay too much stress on the adverse or problematical aspects inherent in it. He presented this new science as an instrument still serving those aims that traditional culture, inspired by antiquity, had always cherished. In sum, science was a neutral tool on the ethical plane, apt for good as well as for evil, Aesop's tongue.

When one looks a little closer one realizes that Bacon foresaw that knowledge, in becoming power, was bound to raise certain problems. It is enough to look through the *New Atlantis* to recognise this. We read there, for example, that scientists will reserve the right to decide which discoveries can be revealed to the state and which must be concealed from it. But Bacon did not question what would happen if the state, in its turn, regarded some discoveries as too important to be made public and science itself too important to be left in the hands of scientists.

In the eighteenth century such a question would have seemed absurd and in fact, the *credo* of science is affirmed in the Charter of the Royal Society, which sets it as the aim of perfecting the knowledge of natural things and of all useful arts, "not meddling with Rhetorick, Metaphysicks, Morale, Politicks etc." Politics is excluded as it interferes with the scientific approach. And yet it is already implicit in the association of knowledge with power.

The apolitical character of the intellectual approach does not rule out the idea of a contract with the State under which scientific societies would be patronised and subsidised by the State, but on condition of being independent of it. Thus, a tension is set up, a threat hangs over the autonomy of the approach arising out of the dependence of the institution on the State. But this tension did not become serious before our present century for at least two reasons: first, little support was required for research activities, and secondly, and above all, science impinged but little on society.

In our day, the tension has become dramatic. Science desires to confine itself to pure reason, but the sciences have become a social institution, and such institutions are not necessarily governed by pure reason. In the eyes of governments, science is a national asset, a decisive factor in the balance of power, and an indispensable tool in the exercise of government itself. Scientists can no longer profess their indifference to the political use which is made of their discoveries.

Until recently, military research was content to adapt civil technologies to the needs of war, and this required no radical innovation either for science or for politics. Then during the Second World War, scientific research was used for the first time as a source of new technologies whose influence could be decisive not only on the conflict but also on the whole postwar period. The authorities, therefore, could no longer leave science to itself; they had, on the contrary, to force the pace of discovery and innovation — and science, mobilised in wartime, continues to be mobilised in time of peace. The mobilisation of science is the subject of permanent arrangements, scientific research is organized, co-ordinated and even planned by governments. That is why the return to the exclusive service of research, of which Oppenheimer dreamed, is a fantasy, a nostalgia for a bygone age; history has burst into the serenity of the laboratories and is there to stay.

Governments can no longer do without science if they want to meet their needs, which have been multiplied, enlarged or created by the development of scientific techniques. They must include scientists in the formulation and execution of policy, as counsellors, administrators, diplomats and strategists. If science, for its part, sets up at the heart of politics, it is because it, too, cannot do without the State.

There are no longer any Maecenas or private foundation at hand to bear the cost of the capital investments in needs, which, by definition, are infinite. The change of scale which started with the Second World War has placed science in a position of growing dependence on government; policy *for* science has become inseparable from policy *through* science.

An age of science which could be called classical, in the sense that its values were imposed exclusively by reference to truth, has closed with the promises of rapid applications which it is now in a position to keep. Scientific research pays no less dear than other activities for the links it maintains with the industrial system. Abundance has its counterpart in organisation, programming and planning.

The quantitative change of which research activities have been the theatre since the Second World War has been reflected in a qualitative change; from an ideal, a vocation, a culture limited to a chosen few, science has become a mass profession; as the source of innovations which can be rapidly exploited, it forms an integral part of the production system. In the eyes of some scientists, this change seems like a betrayal of the ends (and the interests) of science, in so far as it should be concerned solely with the pursuit of truth. The horizon of utility under which it has blossomed compromises it, alienates it and, in a word, prostitutes it.

I have suggested the term "technonature" for the area in which the interests and attitudes of scientists are inextricably bound up with the authorities. Why technonature? The word recalls Galbraith's technostructure, and this is no accident. Galbraith is referring to the holders of technical information and knowledge who participate in the decisions of the modern industrial system founded on the organized use of capital and technology. But Galbraith is careful to distinguish in this community "the corps of educators and scientists" from the others, and he bases this distinction on their motivations; scientists are not seeking power for the sake of power, or fortune for the sake of fortune.

Between this somewhat idealistic concept of the role played by scientists in the industrial system and that which detects a sort of conspiracy among the scientists associated with the "military-industrial complex" it seems to me that some qualifications should be drawn; to speak of a conspiracy implies that there is a group, a consciously organized faction, wanting to seize power for its own profit. To speak of scientists as a breed immune from the temptations of power and fortune is to give too much weight to fine feelings!

In reality, the nature of modern science and the structure of the industrial system combine to make scientists and politicians inevitable partners. There is no need to think up a conspiracy to see that scientific research depends on the objectives pursued by a government, and it is a highly romantic view to attribute to scientists more disinterested intentions than those of other mortals. No doubt the quest for power, or fortune, is not their main goal, but it so happens that, by their function, they also encounter in the course of their career power and, sometimes, fortune.

Atomic research is the most striking example of the power of technonature to generate a new situation out of a marriage of science with the interest of the State. But it is not unique. All fields of scientific research from the atom to space, from the science of materials to the life sciences, from the most theoretical calculus to the behavioural sciences now fall within the scope of technonature.

Let us be quite clear: this institutional dependence of science on the State does not mean that the authorities can more easily or more effectively influence the content of the scientific discourse. We are no longer in the days of Galileo or at least, while there may always be cases of this kind—one thinks of Vavilov or Medvedev or Amalrik—the scientific institution today is strong enough and sufficiently well organized and recognized internationally to unseat all the judges of the new Galileos.

In no part of the world can the political authorities dictate to science its procedures, the laws of its activity, its substance; it is not within the powers of governments to determine the form or the content of the scientific approach. The road to truth is as immune from political decision as truth itself; at a deeper level, truth—whether it be scientific or not, but *a fortiori* if it is scientific—has its own authority which cannot be vitiated by the authority of the powers that be. The public authorities can possibly restrain its exercise by preventing its access to public discussion, by screening part of its results or by distorting their meaning, but there is no constraint or persuasion which can change what it has established, except the authority of the scientific discourse itself.

Even though the political authorities cannot tell scientists how they should search or, *a fortiori,* what they should find, everywhere they aspire to tell them what they should look for, inasmuch as they guide scientific manpower into this or that field of science rather than another. In our day, the despotism of truth is less frequently challenged than its application in the fields over which it exercises its jurisdiction; the characteristic of the new relationship between science and politics is that their possible conflict lies no longer only on the plane of truth, but also on that of performance.

It will be said that the latter plane is not one on which pure research should be called upon to render an account of itself. But how is it possible to isolate the type of scientific research whose contact is immune from all pressure on the part of society? The answer would be easy if a strict frontier could be drawn between the territory of pure research and the territory of applied research. But that is the difficulty which has been made insuperable by the changes that have occurred in the structure and status of scientific research.

Pure science is now no more than one element among others in the system constituted by research activities. Apart from pure mathematics —and even that is doubtful!—it is hard to say where the fundamental phase of research begins and ends. In reality all contemporary research is made up of a coming and going between concept and application, between theory and practice, between what Bachelard called "the spirit which works" and "the matter which is worked".[3]

In antiquity the distinction between theory and practice was a matter of metaphysics, in our day it is merely a matter of psycho-sociology. But if the motivations and objectives of scientists do in fact follow a dividing line, this has no meaning except for scientists themselves; from the point of view of society and even more of the authorities, this purely subjective frontier confers no privilege of immunity.

Science and Scientists' Responsibilities

The ideology of science proclaims the autonomy of research, just as the consumer is proclaimed sovereign in the market economy. But in practice, just as the market economy does not function freely, so research activities do not develop independently of social demands and pressures. Willy-nilly the scientist takes sides when he agrees to carry out a programme which depends on state support, and, at the same time, he leaves the sanctuary of the serene certainties of research to face the uncertainties of action.

The neutral discourse of science declines all responsibility for the use made of the results of research, but it is in fact in the light of these results that science finds its social legitimacy; it contests, in the name of disinterested purpose, all servitude to the economic demands of the modern industrial system. It is in fact because it is thought of in the short term as a technique that it finds itself allotted priority in public investments. It refrains from decisions affecting the fate of nations, but it is in fact because its objectives, quite as much as its results, form part of the setting of the political scene, that it can make its voice heard there, if only to proclaim its right to state support.

The end of *laissez-faire* in the relations between science and politics means the emergence of scientists as an ambiguous species of the genus of political animals. An ambiguous species because it always denies belonging to that genus. And yet, science, looked at as a social institution, is in no way the apolitical sanctuary in the shelter of which scientists believe or hope or profess that they can protect their discourse.

The Oppenheimer case shows how the scientist who proclaims an idea of science outside the field of politics is caught in the trap of responsibility and condemned to an inescapable ambiguity; the old antagonism between knowledge and power, placed under the auspices of truth, now leads to a personal choice between the impossible mission of accomplishment without the support of the authorities and the compromising of that mission by the support it receives.

The scientist who questions himself about the consequences of his work cannot evade his responsibility. He can do so only by proclaiming that the theoretical function of science has nothing to do with its practical function. "It's nothing to do with me, it's society's fault" is then the convenient formula, the salve to the conscience of the Pharisee who delegates his share of responsibility to others. If one can joke on this subject, it is rather what is expressed by Boris Vian's handyman, when he discovers, in *La Java des Bombes Atomiques*[4] that it is not enough to search, or even to find, since the real problems only begin after.

For years and years I've racked my brains
To give my bombs a longer range,
And wholly failed to understand
That only one thing matters—where they land.

How can these problems be disregarded? Think, for example, of over-population. It is not enough to solve it by perfecting the contraceptive pill. The public must also be initiated in its use, especially those for whom over-population is a question of life and death. A great deal was said yesterday about molecular biology and the potential threat from controlling genes; here again, evil as well as good may come of progress. It would be too convenient for scientists working in this field to shut themselves up in their laboratories for the pleasure of doing research without recognising that, by their very research work, they are both producers and consumers of social change; consumers, because, after all, they depend on the wealth granted to them by the community; producers, because they transform this wealth into discoveries which are factors of change. On this point, at least, Marx was right; knowledge has become a factor of production, and if it has done so, it is because it is scientific knowledge.

Does that mean that the responsibility of the scientist gives him special competence in the political field? Most certainly not. The scientist is not an expert in politics; his competence in his own field gives him no greater authority in other fields than any form of technical skill. Objectivity is not transferred to human affairs by the mere fact of applying scientific methods to them.

Among scientists who make pronouncements on political problems there is often a certain naïveté in thinking that value judgments and ideological options can be reduced to clear and precise terms. In a political combat the scientist is no less partisan than the militant who has no access to scientific language. The special characteristic of positivism is to imagine that it is possible to work out an applicable operational technique in the presence of conflicts and opposing values. But it is clear that, whatever the progress of the mathematics of decision, they will never succeed in formalizing the data and the choices of the political universe.

The scientist who intervenes in urban politics is no better armed than any other citizen to settle the problems of the city. But he is at least better armed to throw light on the problems connected with the role of science in the city. Here lies the whole foundation of his social responsibility. In this capacity, and on pain of complicity, he can and must effectively carry weight in political decisions.

So long as theoretical activity could be divorced from its applications the scientist owed service solely to scientific truth. Today, when knowledge can no longer be distinguished from its consequences, the ethics of his profession demand from the scientist a new duty, namely that of informing society of the implications of what he is looking for and what he finds. There can be no science without conscience.

If it is true that modern technology is wholly dependent on science, then it is the function of the scientist to be the conscience of technology. To inform, to educate, to warn. In the name of the ideology of science conceived as the discourse of truth, the scientist is duty bound to denounce and fight what he conceives to be the misuse of science. It is only at this price that the priests officiating in the cathedral of science will cease to be confused with the merchants in the temple.

Cathedrals bear witness not only to the anonymous workers who built them, but also, and indeed primarily, to the faith they are designed to shelter. Was it not one of the organizers of this meeting who said in accepting the Nobel Prize, "Scientific research is a religion which calls for faith, a rational faith. Like every religion, it must have its prophets, a college of apostles and the heart and soul of a whole people. It must also have its martyrs."[5] But one could also add that it includes, like all religions, its share of magicians, simonists and time-servers.

If it is true that faith is compromised by the merchants in the temple, then the priests of science should dissociate themselves from those who reduce the practice of science to a purely utilitarian adventure, to works and objectives in which it can no longer recognize itself. For, if science is a sort of faith, it is so because first and foremost it stakes everything on the power of reason to render account of the world and to mould it to human ends; unless scientists assume full responsibility the whole institution of rational knowledge will finally be compromised.

The end of *laissez-faire* in the relations between science and politics cannot mean there is now unlimited freedom to innovate; it is a matter for scientists themselves to contribute to the control of technical innovation, that is to say, to bring weight to bear on political institutions to ensure that the criteria for the support and exploitation of science are not based solely on output, profit, and short term prospects.

Otherwise, by dint of being associated with the merchants in the temple—most of whom, moreover, are nowadays arms merchants—scientists will be implicated in the erosion of the faith they profess, and the disappearance of the scholar, who is the conscience of science, will deprive the cathedral of all human meaning.

In conclusion may I say that the fact you are meeting here to discuss "The Impact of Science on Society" proves that there is no reason to despair.

---

1 A. Hunter Duprée, *Science in the Federal Government,* 1957 re-issued by Harper Torchbooks, New York, 1964, p. 375.
2 Cited by Karl Jaspers, *La Bombe Atomique et l'Avenir de L'Homme,* Buchet-Chassel, Paris, 1963, p. 360.
3 Gaston Bachelard, *L'Activité Rationaliste de la Physique Contemporaine,* P.U.F., Paris, 1965, p. 3.
4 Boris Vian, *Textes et Chansons,* Collection 10/18, Paris, 1970, p. 39.
5 André Lwoff, Speech of Thanks, Stockholm Town Hall, December 10, 1965.

Professor Michael Feldman is Dean of the Feinberg Graduate School of the Weizmann Institute. He was born in Tel Aviv on January 21, 1926 and studied biology at the Hebrew University in Jerusalem. From 1953 to 1955 he was British Council Scholar at the Institute of Animal Genetics at Edinburgh University. He has been a Visiting Scientist at the University of California and the National Institute of Health, Bethesda, Maryland, U.S.A. and a Visiting Professor at Stanford University, California. His research has been concerned with immunology and developmental biology and he is Head of the Department of Cell Biology at the Weizmann Institute.

Science and the Crisis of Democracy

Prof. Michael Feldman,
The Weizmann Institute of Science, Rehovot

Many prophets of social development assure us that science and technology will trigger profound changes in the social and political structure of western societies. In fact, all those who do so are engaged in prophesying the past. Science has already affected the social and political foundations of the western world. Yet, the effects have not been universally realized, let alone appreciated. My intention here is to outline certain aspects related to the impact of modern science on the political structure of western society with regard to what I consider the present, and even more the future crisis of democracy.

The recent history of the western world did not follow the predictions made by great scholars of society in the 19th century. Karl Marx claimed that the capitalistic system would collapse even without the direct intervention of socialistic movements. He suggested that the rhythmic crises of capitalist economy would increase in both frequency and intensity, as a result of which the poverty of industrial workers would increase. This in turn would result in intensification of the struggle between industrial employers and employees, i.e. of the class struggle, leading to disintegration of the whole capitalistic system.

Reality, however, did not confirm Marx's predictions. In fact, the frequency of economic crises has decreased, leading to a complete abolishment of their rhythmic nature. The standard of living of industrial workers has not decreased but has increased progressively. Industrial workers no longer constitute a proletariat, but rather a community of skilled and professional workers. As a result, the struggle between classes has been decreasing and in countries in which technology has reached a high level, the class struggle has practically been abolished. Although the differential between rich and poor has not changed, the mere increase of the economic zero-point, i.e. the income of workers in industrial societies, has created a state of social stabilization.

Marx had not seen this trend of development. At that time neither he, nor any other scholar, could have predicted that the sciences would undergo a dramatic development, thus creating a basis for technological progress which became the foundation of the new industrial revolution. The outburst of production of modern industry enabled and in fact necessitated a significant increase of the purchasing power of the masses, which was achieved by raising the income of industrial and other workers.

Science and technology are thus responsible for the economic progress of the working classes in industrial societies, and probably more so than are the socialistic movements which have aimed at the same goal. Once income of workers increased, the appetite for a social and economic revolution decreased. (The revolutionary tendencies manifested lately by students and intellectuals seem to hardly touch the working classes.)

One of the most significant consequences of the abolition of class struggle in industrial states has been the disappearance of ideological differences between the alternative political powers. Let us consider the political arena in Britain at the beginning of this century. The two parties were the Conservatives and the Liberals. The basis of these political organizations was ideological in the sense that each of them sought the materialization of a different social and economic structure in Britain. These differences stemmed from profound differences regarding ethical values, social principles and economic goals.

Are there such differences today between the two parties in Britain? Do the Conservatives today try to implement a social structure which is factually different from that which the Labour Party advocates? There are today hardly any concrete ideological differences between the Labour and the Conservative Parties in Britain, as there are hardly any ideological differences between the Democrats and Republicans in the United States. Technology, by creating industry, has led to economic progress which has brought political stabilization to the industrial societies, and this in turn has deprived the political parties of ideological dogmas.

I wish to stress one point: I do not claim that ideology as such has no place in today's western world. What I am saying is that ideology is no longer the basis for political organizations in western technological societies. Of course the situation is different in those states in which industry lags behind. Hence, political leaders in industrial states no longer rely on ideological doctrines to determine their attitude towards any important issue. And this state of affairs carries the seeds of what appears to me as the coming crisis of democracy.

Until recently, i.e. during the classical era of western democracy, ideology determined the attitude of politicians toward the major political, social and economic issues. During those days one could have predicted what would be the view of a given politician, either from the left or the right side of the political spectrum, towards any of the important issues. Thus, one could have predicted the political decisions of, say, Léon Blum in France, since these were to a great extent dictated by the ideological principles of his socialistic party, which were known to everybody. Hence, in those days, when the individual elected a party to office, he elected, in fact, the decisions he wanted implemented.

On the other hand, today, when there is no longer a direct relationship between political ideology and actual decision-making, the individual elects delegates whose decisions he cannot predict. Inevitably, the politicians often make decisions to which the people who elected them object. The individual citizen can obviously express his objection in public, yet this hardly affects the decision-making apparatus.

The loss of predictability regarding decisions on public issues, due to the loss of the ideological basis of the political parties, is a new phenomenon with severe socio-psychological implications. In fact, these implications seem to constitute the very origin of the unrest of students and intellectuals in western societies. The young individuals realize that their capacity to determine the decision of their representatives is exceedingly small. Thus, technological progress, leading to economic achievements, has undermined the ideological basis of the political bodies. The ability of the individual to affect decision-making by electing his representative has been dramatically diminished. It is only natural that he should revolt against it.

It is generally accepted that during the last 50 years profound changes have taken place in every avenue of life. The economic system changed, the social structure changed, industry, engineering, art, music, poetry — all basic components of western society underwent dramatic changes. On the other hand, one important element of our life has retained its original pattern and was hardly affected by the general transformation in western culture: the structure of our democratic decision-making apparatus.

The original shape of parliamentarianism, which constitutes the basis of our concept of democracy, has remained almost unchanged since its early days. The individual's revolt against the system of decision-making which no longer echoes the individual's views seems to lead to a crisis, because there is no simple democratic alternative to the present system. We know what is wrong, yet we do not know how to substitute for it a better political operation.

There is obviously a simple solution to the justified desire of the individual to participate in political and economic decision-making. One could suggest that instead of letting his representative decide, let individuals themselves express their view on every important issue. We could then register the view of the real majority of the citizens, rather than that of politicians who no longer truly echo the views of the electorate. This, however, leads to some difficulties inherent in the most elementary principles of our democratic system.

When democracy was conceived, it was not based on the assumption that we can rely on a majority that will make the right decision. Rather, the concept was, that by definition the *decision of the majority is the right decision*. Indeed, as long as decisions were essentially a choice between alternative ideologies, i.e. between ethical and social values, this concept was unchallengeable.

Today the choice is no longer between general political doctrines, but rather between alternative means, methods, techniques for achieving the same general goals. These are based on detailed scientific and technical knowledge. Does then the principle of confidence in the choice of the majority still hold in its original form with regard to the crucial problems which face modern society? I raise this question though there is no doubt in my mind that there is no alternative to the democratic system. Yet, it does seem that democracy should revise its operational tools for decision making.

The recent history of Western Europe can be divided into three distinct eras: (a) An era of struggle for equality in civil rights, which in fact terminated the acquisition of more, rather than actual equal rights. (b) An era of struggle for the equal distribution of capital, which again terminated by the acquisition of more, rather than equal capital. (c) An era just starting, in which the struggle for equality in know-how will characterize the social stress. Know-how, rather than rights or capital, will determine successful leadership in a technological world.

Know-how has not been an important parameter in determining the achievements of political leaders in the past. In England, for example, the majority of Prime Ministers have come from the universities of Oxford and Cambridge. In these institutions they had been exposed to an educational system in which training towards a broad and gentlemanly outlook on world affairs was stressed. At periods, when decisions were based on a choice between alternative ideologies, this education was sufficient, if not always the best preparation for leadership.

When, however, decision had to be based on choice, based on deep complicated and sophisticated know-how, that type of leadership could hardly cope with the new reality. And anyone who wishes examples should read the 20th century history of the Middle East. The emerging problem today is not just how to prepare the leadership for proper decisions in a scientific era, but how to make the individual citizens participate intelligently in decision-making based on know-how.

We all wish that political choice be founded on firm ethical values. Unfortunately there are no objective criteria for ethical values. As a result science, as such, is indifferent to ethics. Yet, in the everyday operation of science there are two principles which seem to me to manifest ethical values: Firstly, science develops in a non-authoritative way. The scientific truth is recognized not according to who expressed it, but by what has been expressed. A Ph.D. student can make a scientific discovery which contradicts the truths accepted by all important authorities in a particular field. Yet, the moment he has proved his discovery, all previous truths no longer exist, no matter how great and authoritative were those who believed in them. There is no other field of human activity in which such a non-authoritative reality prevails.

Secondly, science develops through scientists communicating to their colleagues the whole truth about their discoveries. Until a scientist does that his claim cannot be verified by others, and unless it is verified his claim has no impact on further scientific development. There is no other human activity in which progress is determined by the exposure of the whole truth.

Thus, science makes progress while operating on two principles which seem to me to embody important ethical values. However, this is not a result of free choice of moral principles. Rather, it is a result of the very elementary basis of the scientific methodology imposed on the scientists' behaviour, not because they choose to behave ethically, but because they choose to do scientific research. The lack of authoritative behaviour and the exposure of the entire truth applies only to their scientific work. What then will ensure a humanistic or ethical consideration in decision-making with regard to all other components of our life in an era in which ethical doctrines (ideologies) no longer furnish the basis of political operations?

Had man been born in the image of God, we would not need to worry too much. Great scholars like Marx and Freud seem to have believed that man was born in the image of God. Marx believed we only had to change the economic structure of human society for the godly attributes of men to become apparent. If we alleviate some of our sexual repressions, thought Freud, the image of God would be revealed in man. Both Marx and Freud had essentially a romantic view of man.

In reality, human behaviour, both at the individual and social level, seems to be determined by genes which may be inherited from our prehuman ancestors and which need not elicit in us any feeling of pride.

Australopithecus africanus seems to represent the last ape before man. Raymond Dart, the anthropologist who discovered it, found evidence that it was a killer, that it went armed. The editor of the journal in which Dart published his views on the behavioural properties of our ancestors stated, in an appendix to the paper, that "The conclusions of this paper are relevant to the African ape. With regard to humans, they may be relevant to the bushman and other human tribes in Africa, not to us". Yet, we all know the truth.

The ethical choice in a scientific world is therefore an open, yet pressing problem, for this choice might very well have to operate against some of our genetic properties.

Discussion

Raymond Petit, Press Officer, Scientific Youth of Belgium, Brussels

I have just heard Professor Salomon paint an extremely sombre picture of the gulf being created between society and science. As a representative of the Scientific Youth of Belgium, an association dedicated to encouraging boys and girls in grammar and high schools to opt for careers in science, I should like to point out that there are people who by their work and enthusiasm are doing a very great deal to bridge this gulf.

We work with about 8,000 young people in some 150 sections. In each section a teacher of science, a chemist, a physicist, a biologist, sometimes even a historian or other enthusiast, catalyses and tries to direct the overflowing enthusiasm encountered. By our contacts within the universities we can almost better than anyone else orientate the desire for the fulfilment of a career in the sciences. We can, for example, say to a boy or a girl, "No. The field you have chosen could between now and the time you leave university come up against a wall of under-employment, or against such competition that you will not find a job in the place you want."

Equally, with the help of the universities and the technical colleges possessing the materials and the necessary equipment, we can organize courses of instruction. We organize them for our members who are at most 18–20 years old. In other words they go to university to do courses in chemistry, in physics, and for the first time, the Free University of Brussels is to open a course where they can handle a computer. This had not been done before.

There are, I might add, at the moment more than 80 to 85 nations of the world who have got together organisations very like the Scientific Youth and which are doing well.

You will understand that we need university aid. We need, and we already receive, the aid of private firms such as Shell and Philips. All of this adds up to the fact that our young people on leaving secondary schools are better informed about what they are going to do at university. By this single fact they gain at least several months' advantage which will permit the universities and our economy, in short our society, to benefit all the sooner from better men of science.

Prof. K. Hozelitz, Mullard Research Laboratories, England

I would like the physicists to make a plan outlining possible limits for future research.

Professor Victor F. Weisskopf has been Professor of Physics at the Massachusetts Institute of Technology since 1946 except for the period 1961–65 when he was Director General of CERN in Geneva. He was born in Vienna on September 19, 1908, and educated at Göttingen, receiving his Ph.D. in 1931. He became a Research Assistant at Berlin University for a year and then joined the Swiss Federal Institute of Technology. In 1936 he was at the Institute for Theoretical Physics in Copenhagen and the next year became Assistant Professor of Physics at the University of Rochester in the U.S.A. He served as a Division Leader with the Manhattan Project from 1943 to 1945. Professor Weisskopf has received honorary degrees from nine universities and is the author of *Theoretical Nuclear Physics* published in 1952.

Conclusion

Prof. Victor F. Weisskopf, M.I.T. Cambridge, Mass.

I am in a difficult position, for I am supposed to summarize this meeting, and you will perhaps bear with me when I say that this is practically impossible. All that I can do is to tell you what I think about the problem.

You have heard very contradictory statements, and this is of course not surprising. These contradictory views about science and its relation to society show how centrally located science is, and how tremendous is its significance for our life. Something that penetrates every part of our thinking, of our acting, of our suffering, such a thing reflects all the aspects of the human condition. These aspects are numerous, tragic, wonderful and everything else at the same time.

Somehow I believe that science has developed from a state of youth to a state of adulthood. When you are young you have great ideals and you think the world is wonderful and what you do is wonderful, and you will change the world. When you get old you see that this is not so easy, and that when you try to change the world, you usually change it for the worse; in fact you cannot change it. This, I think, is our present state of affairs; science has become adult: I am not sure whether scientists have.

Science, definitely, is on the defensive these days. This is a relatively new phenomenon. The melodies you would have heard had this meeting taken place six or seven years ago would have been quite different.

Essentially there are two attacks on science. One is that science is an expensive luxury, which should be supported by the public only if there is a promise of immediate pay-off in terms of practical applications for industry, medicine or national defence, but not as a study of nature for nature's own sake, especially when it is so expensive, since this is not of great relevance and has no public value.

The other point of view is that science should be severely curtailed (of course both points of view are anti-science) because it is the source of industrial innovations, most of which have led, and will lead even further, to the deterioration of our environment, to an inhuman, computerized way of life. Science will destroy the social fabric of our society, lead to more dangerous and destructive weapons, to annihilation in the next war and to far-fetched developments in our society as foreshadowed in Orwell's world of 1984. The best science, according to this point of view, is a waste of resources that should be devoted to some immediate, socially useful, purpose.

It is actually easier to deal with the first point, because the second point of view contains a lot of truth, and it is, in fact, in some ways a justified point of view, if you look around and see the abuses of technology. And though I do not want to refute the facts, I may want to refute the conclusions.

As to the first point of view, I do not think I should spend much time on it. Basic science, science for its own sake, has done an extreme amount of service to industry, to medicine and so on, as you all know. I should perhaps only say one thing about the expensive luxury. You can easily calculate the total outlay for basic science—from Aristotle to Sabin—has cost no more than ten days of United States' production. Is it then so expensive? But this is, of course, not really the point.

Conclusion 195

The second point of view, to which I would like to address myself at greater length, claims that science has produced technology, and technology has produced all these terrible effects; hence, we must stop science in order to stop technology and its terrifying consequences. This is the famous moratorium argument, which has been discussed often here.

My reaction, and the view that I suppose scientists in general should take, or maybe everybody, is that it is not a question of *stopping* science and technology; it is a question of *changing* it. If you have built a house, and you find that you have done it not too well, you try to build another one, but you do not throw away the tools. This is really what the supporters of a moratorium propose.

What is wrong with the house that was built, what should be changed? This could be the topic of a very long talk. Professor Salomon has mentioned some aspects of this subject, and Katchalsky and many other speakers have also talked about it. Let me put it another way: the error in respect to the environment question is a matter of wrong cost accounting in the widest sense.

We have not accounted for the long range effects, partly because we have not paid attention to them, and partly because we lack the necessary knowledge to do so. There are of course consequences of technological development that go much deeper and are harder to formulate than the deterioration of the environment. Because it went so fast, it created deep social consequences which we are not even able to describe and formulate correctly.

People can no longer live in the present situation despite the fact that they seemingly live better the more they consume. But just one look at the cities, particularly in America, will convince everyone that something is very, very wrong. Some of these problems could have purely technical, and therefore scientific aspects but probably to a much greater extent they are social problems. The latter are much more difficult because we have no experience with them; we should have, we could have—but we don't have. They are more difficult than any problems mankind has faced.

I am not enough of an historian to say whether similar situations have happened before; they probably have. But I am not sure if we can learn from the experience of the past since we are facing a completely new situation, namely a situation of exponential growth, in particular in population.

We have discussed the problem how to stop the population explosion, how to go over to a stationary society with all its economic, social, psychological problems. Psychological because what will people live and strive for if the situation is stationary? What will be their ideals, will they be bored, will they start fighting each other?

This is a problem we will have to face because we cannot go on growing as we did. But this, of course, is not the job of the natural scientist, as most of us here are. We can only point towards the problems, not much better than anybody else can, for we are laymen. I hope fervently that those people who call this their profession know more about it.

The technical side, however, is not to be neglected. There are many problems, especially those connected with environment, transportation, and even city life, that are at the centre of our crisis, and to which natural scientists can contribute a lot. Many people say that because these problems are so terribly "burning", science should devote all its activities to

Conclusion

them, and we should all become, for example, environmentalists. Now, there is a great danger in this.

Let me read to you a quotation by Michael Polanyi from his book *Personal Knowledge*, in which he speaks about the scientific method: "The scientific method was devised precisely for the purpose of elucidating the nature of things under more carefully controlled conditions and by more rigorous criteria than are present in situations created by practical problems. These conditions and criteria can be discovered only by taking a purely scientific interest in the matter, which again can exist only in minds educated in the appreciation of scientific value. Such sensibility cannot be switched on, at will, for purposes alien to its inherent passion."

In other words, he says that to solve the natural science problems in our environment in its widest sense, we need scientists trained in pure science. If we should take all the scientists now working in pure science and put them to work solving these problems we would not have the young people to take their place. We need, therefore, a vigorous establishment which practises basic science, science for its own sake, science to find the reasons and causes of things, to follow up the connections in nature, to determine the laws of nature, for this is how young scientists get the training, the attitude, the spirit of research necessary to solve so many problems. And you cannot replace this, as Polanyi has said.

What I want to stress here is not so much the results, but the attitude, the spirit, the state of mind which is created in the practice of basic research which is the basis of our ability to deal with nature, which we must not destroy. Our universities should not turn out environmentalists. Rather we must train geologists, physicists, chemists; we must train biologists in the basic sciences so that they can apply the skill, the attitude, the state of mind in which they were trained to those problems. Only then will we succeed.

I will hasten to add that the value of basic science does not reside exclusively in those points which I have mentioned. It is very much deeper. And I should also say that the attack on science today also goes deeper. Let me first speak about what I consider the value of science. I believe it lies in the effort on an international level to gain insight into the workings of nature, insights that have the tremendous effect of producing the power which we now have, fortunately or unfortunately, to influence nature.

Somebody said that natural science is really a western product. I do not think this is so. Perhaps it had its beginnings by accident in the Western World, but if you watch how the Japanese and the Chinese, who are not connected with us at all, are working, producing, creating, contributing to this collective effort, you will see that its great value lies in bringing people together.

The last few decades have given us insight into what nature is, how the universe developed, and what is the basis of life. As you all know, these discoveries are equally as great as the discoveries of the 19th century about the nature of electricity, or those at the beginning of the 20th century about the nature of matter. Why is it then that these discoveries are feared, despised, attacked? It is not only because they produce pollution, and give great power to those who abuse that power; it is deeper than that. It seems that people regard the natural sciences as something alien, something far removed from what they think the relation of man to nature should be. This is a very common way to look at science, but for most of us to whom science is a revelation of

Conclusion

the mystery of things it is a great disappointment. Why is it a cold, alienating element for others?

I think there are two reasons, and I think the scientists are to be blamed for both. One is the arrogance of science in regard to its universality. We say, and with a certain justification, that science covers everything. Not that we understand everything, but we claim that science has the ability to push its frontiers to where it will eventually understand everything. Our past experience shows that this is to a certain extent true. But we have forgotten something, and I think this has something to do with the attitude of the public towards science. We have forgotten that science is universal, but it does not transcend everything.

To give you a very trivial example, take a Beethoven sonata. According to physical science a Beethoven sonata can be just a line on the phonograph record, complicated, but a line. To a physiologist it is probably a line of electrical impulses in the brain. But obviously, neither of these is *the* Beethoven sonata. The content of the Beethoven sonata transcends the physical description, even though everything that is of the Beethoven sonata is included in the physical description. In some ways, therefore, science is universal, but it is one-sided. It leaves out quite definite values. I use the word "values", although I am not quite sure what term to use, but probably values are among those things that this description completely leaves out. It may perhaps sound trivial to say science is universal but one-sided, but it is often forgotten, and I think it is a very important point. I will not grade these things by saying that what science leaves out is more important than what it encompasses. It is just another aspect of this world, of the human condition.

The second point where I think we are at fault is the distortion of science by the scientific establishment, by the tremendous development that has taken place in the last twenty years (some people would say by the tremendous amount of money that science got, and they may not be altogether wrong). People have forgotten the "mystery of things", and they just look at science as an organisation for producing new results. With this goes all the over-specialization. As Professor Scholem so discerningly pointed out, most of the so-called scientists are not really scientists in the true sense, but they do constitute the majority of the scientifically trained. And in the science establishment the philosophical idea, the mystery of things has become lost.

Science has become complicated; it requires one to be a virtuoso in its technical or mathematical aspects, or both. People have forgotten that this virtuosity has a purpose, and the purpose is to find the laws. They are simple, but they are subtle. It is this simplicity and subtlety that are the mark of science, as Einstein has demonstrated in a somewhat different connection. It is therefore the responsibility of scientists to bring this out in education and elsewhere. They should make it clear that science is not a cold world of numbers, but on the contrary creates a greater intimacy with nature, with the universe. By doing this they will demonstrate a greater sense of responsibility, and generate an awareness of the unique world in which we live, thereby denying the concept of alienation, quantification and dehumanization.

Let me say a word about the difference between the crisis in Europe and in the United States. In the United States it is definitely much deeper and more tragic than it is in Europe. Why is it that in America where there is more science, and therefore more technology, there is more trouble? Europe, it is true, has an older tradition and places more value

Conclusion

on contemplation and perhaps on beauty. But it is the general observation that when things go up they go higher in America, and when things fall, they fall farther than in Europe. The difference stems to a large degree from the quality of education in the secondary schools.

In Europe, fortunately, secondary school teachers receive a much better education than those in America. There is a demand everywhere for good science teachers, and it should be remembered that not everybody who goes into science needs to take up research. It is terribly unfortunate that if you want to count in science, you must undertake original research, for this is devaluating original research, and many people make no worthwhile contribution. They would do much better if they would apply their talents to interpreting science, which is in fact a lot more difficult. It is this kind of trained talent we need in the schools.

Let me just end up with a small summary. I have talked a lot about basic science, but as somebody has said today, basic and applied science belong together, you cannot separate them. Science has to be considered as a whole. The present trends are against basic science and in favour of applied science. They are shifting the equilibrium, and can destroy the fabric of science. Let me illustrate this by likening science to a tree, the basic sciences forming the trunk, the older being at the base, and the newer, more esoteric ones near the top where new growth takes place. The branches represent the applied activities which emerge from all the basic sciences — the higher, smaller branches being the outgrowth of more recent basic research. The top of the trunk, representing the frontier of basic science, has not yet developed any branches.

In applying this picture to physical science, for example, you would put classical physics, electrodynamics and thermophysics at the base, and atomic physics a little higher, with its well developed branches such as chemistry, and material science and solid state physics. Still higher you would find nuclear physics with its younger branches symbolising radioactivity, tracer methods, and so forth, and at the top, without branches yet, you would find modern particle physics, and some of the pioneer sciences. There was a time, sixty years ago, when atomic physics was the branchless top.

If you have a tree you have to care for it; it should not grow too fast or too slowly, and you should prune it—but removing only those branches that are no longer healthy. In the example I am using, pruning involves priorities and this is not easy. Though many people would like to, you cannot sever the trunk, leaving only those branches that have the nice red apples. How then do you decide which sciences you cut and which sciences to let grow? This is very difficult, and Professor Hozelitz suggested the physicists draw up a plan. Let me say that they do; in fact they do too much of it.

Every four or five years, at least in the U.S.A., the physicists are supposed to draw up a programme for the next ten years. This is ridiculous, because the essential thing in basic science is the unexpected. And the unexpected is the one thing that is important from the philosophic as well as from the practical point of view. One could and should have priorities for technology, because there is a definite purpose to fulfil. This is difficult as we have seen in the discussion about the SST, but there priorities are dictated by the necessities of life. But we must be wary of definite priorities for science. The main point is to cut out mediocrity, the bad branches, and the present cutting craze in America, if done well, may perhaps be quite useful.

I think I express the ideas of Weizmann and the fundamental ideas of the Weizmann Institute when I say that science cannot develop unless it is pursued for the sake of pure knowledge and insight. It will never survive unless it is used intensely and wisely for the betterment of humanity. There are two powerful elements in human existence: curiosity and compassion. Curiosity without compassion is inhuman, and compassion without curiosity is ineffectual.

Prof. Albert B. Sabin, Rehovot

In the last two days we have had the extraordinary experience of seeing what amounts to nine outstanding scientists in different fields of activity on a psychiatrist's couch, freely associating about some central concept. Now that Professor Weisskopf has put their thoughts into perspective, I am able to pick out what, perhaps, was the problem that brought all these people to the psychiatrist. If I have to carry away one impression from this Symposium, I would say Weisskopf's remark that science is now on the defensive summarizes it.

Almost all the speakers mentioned some defensive attitude into which scientists seem to have been forced. No one said, and perhaps it does not need saying, how much science and technology have done for mankind. I witnessed here nine people somehow or other defending science, and now I am the tenth. I shall try if possible to reverse this trend. I think that we owe it to society to give some indication of what science can do to help society solve the serious problem it faces, to give it confidence in the future, even though the diagnosis may be very serious and the prognosis very poor.

I listened, and I tried to understand the nature of society's fears. Basically, two fears were expressed. One was the fear society has of science and the other a fear held by scientists themselves. What is it society fears? The fear of what the physicists have created, the power of a total annihilation of this planet, or at least of most of it, and the reduction to barbarism. This we have lived with for so long that people have become used to it. In fact, it seems that the neutralization of the power to destroy has diminished the threat.

So, what are the philosophers, the thinkers, worrying about? They are worrying now about a new threat. They are worrying about the developments that may arise from biological research. They may be justified, because the rate at which biological research is revealing new information can be frightening. Supposing we know how to regulate old age; supposing we know how to regulate growth; supposing we even find out how to break the species barrier — what then? Some people think that man was banished from the Garden of Eden because he ate of the tree of knowledge, but this is not so. Fifteen years ago I gave a lecture entitled "The Biologist's Dream is the Humanitarian's Nightmare", and while preparing it I read Genesis again. I found that man was driven out of the Garden of Eden not because he ate of the tree of knowledge and therefore knew the difference between good and evil, but lest he "... take also of the Tree of Life, and eat, and live forever". Biologists are getting close, and the humanists are getting worried. There is perhaps a reason for this worry, and the question is what to do about it.

What is the other fear, the one held by the scientists? They are afraid of what they call the loss of freedom. What is this loss of freedom? As many of the speakers have revealed, there is a tradition that an individual scientist selects his problem, carries out his work, is accountable only to himself and to his colleagues, who will put him through the

Conclusion   205

fire. Today he is afraid of losing this independence, because society is beginning to ask for help, some dividends from the support it gives him, and he fears the change that will take place. He senses that there is time for a change, that science has grown up, that evolution has progressed to the point where you just cannot go on the old basis of one man – one problem.

And now I would like to ask the question: do all scientists have to be humanists? My answer is "No". One thing that all scientists have in common is a curiosity to learn. Never mind what the target is, and I will not ask a scientist to take on oath that he will pursue only things that are good for man. Neither would I deprive him of support should he refuse to take such an oath. I believe that scientists should have a right to ask whatever questions they please, even if they hate human beings.

There is only one demand that one should make of scientists; that they do good work. Of course, if a scientist hates human beings and does not do good work, he should not receive financial support. We all talk as if scientists have a responsibility to be greater humanists than other human beings. Is everybody a humanist in this world? Again the answer is: "No!" Scientists are just like other people, some of them are humanists and some are not. Those that are humanists want to see that the knowledge they help to produce is used to solve human problems, and they are more aware of what that knowledge can do than others. There is, therefore, a special need for those scientists who are humanists to do something, something I will spell out in my conclusion.

Some of the people who spoke here presented us with challenges that require comment. Cramer started off, and he talked about the challenge of the technological future. I took out just one phrase. He said: "Logarithmic growth of science and technology cannot go on indefinitely, and there is some indications of the end of this logarithmic phase." "Indefinitely" is a long time and I do not care to talk about a time span that is indefinite. I am more interested about the next 10, 20 and 30 years during which the course of our civilisation will be determined.

In my judgment the present problem is not over-production of scientists and technologists but their under-utilization, and there is a great difference. If science and society work together they will find the need is so great that there will certainly be no over-production; rather it will be a case of under-production. Cramer spoke of the "age of planning" and "new asceticism". He said: "The affluent society also has an enormous potential for thinking and planning... integral planning seems to be the only way to escape a crisis caused by a struggle for a life of higher order." This is the challenge that we should meet.

Life of a higher order is not meant only for an elite. The aim is to struggle for a life of a higher order for the vast majority of humanity, two thirds of whom live in utter misery and despair. Here is the challenge that science must present to society, and work together with society to overcome. And how to overcome it? Cramer spoke about the new asceticism as a voluntary renunciation not only of materialistic progress, but also of the capabilities of man to transform the material world. I do not mind seeing those who have already made advances renouncing some material progress in order to give others a chance to catch up. I do not mind calling upon the affluent world to begin to restrict its need to the level to which we hope to raise the rest of the world. But when he also asks us to renounce our ability to transform the material world, that is another matter.

Rather science and technology must call on the world's leaders to show how these disciplines supported by money, planning and good management, can create a better world for more people than ever before. This is the clarion call we must bring to society.

I was not here when Aharon Katzir-Katchalsky talked, so I had to listen to him on the tape in my room. Aharon has the capacity to lift you up in the clouds. He spoke of a scientist's approach to human values, but actually what he meant was the human approach to human values; he just happens to be a scientist. As Professor Scholem commented later, there is no such thing as a scientist's approach to human values, only the human approach. I am sure that Katchalsky deserves much more comment than this, but then he has roused quite a lot of comment during the course of this discussion.

Professor Van Hove said many interesting things, but I want to extract one thought that we should carry away, because it is a path for action. He said that present trends, particularly in physics and astrophysics, require ever more powerful instrumentation, more powerful means and greater expenditure. This development, he stressed, requires the formation of a global science policy by which the world's most advanced research instruments will be planned, constructed and exploited in a concerted way by the best scientists on a world-wide basis. Of course he spoke of a policy which is already in existence, and that has already demonstrated its effectiveness in one field. I think that what he had in mind was the need to extend this on a much larger scale and to many more activities. There is no need for the latest electron microscope on every floor; there is no need for a highly expensive computer for every two laboratories. You have to use good judgment.

Pekeris came next. What I want to pick out from what he said was that there are certain fields in which scientists themselves should take the initiative, fields in which they have vision. To me it means that there are times when scientists should not wait for society to prod them. If I understood him correctly, he made a plea for scientists to take the initiative and organize themselves in groups to achieve certain objectives.

Professor Maaløe presented us with a model of the regulation achieved by evolution in the ribosome factory of the bacterial cell. If we like to have models on which to base generalizations and to learn from the whole evolutionary process in nature, this is a good model. All it shows is that in any organized system, be it the simple bacterium or anything else, organization is an absolute necessity for survival. And if it has any meaning it also means that the freedom of science ultimately will have to lend itself to regulation. The only question is how and by whom it will be regulated, and what will be the system that will bring it to its greatest effectiveness.

David Samuel gave a beautiful example of the way a scientist, or a group of scientists, should deal with a difficult problem. He took a field of scientific research which can be of great concern to humanists, and he showed what needs to be done and how to use the new information in some sort of reasonable, wise way. I think his presentation was a model of what one might hope for; you take a big problem, you divide it into small ones, and different people take on different aspects of the problem, and the total achievement then emerges from the sum total of everyone's work.

Weisskopf mentioned Aristotle. I have admired Aristotle for a long time, and interestingly enough not by reading him. As I remember, the reports of the National Research Council in the U.S.A. had a quotation from Aristotle to the effect that the search for truth (in this case make it scientific research) is easy, because none can miss it wholly; and hard, because none can master it fully. But, the quotation goes on, the grandeur, the ultimate grandeur, comes from the work of all. This is something we must not forget. Professor Kendrew called our attention to the fact that the work of so many scientists is drudgery. What you do yourself now must be done in such a sophisticated way, with such sophisticated instrumentation, broken up into such little pieces, that each little piece can only have meaning when put into its final place. The grandeur is in the vision of how each contributor's little piece will ultimately fit in the overall picture. In other words, there is a perspective even in the purest of pure research.

The assumption that fundamental research stopped during World War II is, I think, misleading. There was probably more fundamental research done during World War II, not only on weapons, but in medicine, industry, and so forth than ever before; only the targets were different. Some of my friends from Germany assure me that during the Nazi period there was no stoppage, or moratorium, on fundamental work. What was missing was contact with the outside world. Fundamental research did not stop, and in spite of what Chaim Pekeris said, there was no moratorium during the war, ever. Of course if you want to define moratorium to include balance, that is different. For example, we now have neither the people nor the means to attack all the problems. So if you want to call working for a balance a moratorium —very good.

Much really good philosophical analysis went into the presentations by Professors Salomon and Feldman, and it would be superfluous for me to comment further on what they said.

To conclude, I am not a blind optimist and I believe the world faces a crisis. To me the crisis lies in the poverty-stricken countries, where in the next 20 years there will be more than five thousand million people with not enough to eat, no jobs, and nothing to look forward to. This is the crisis which I think will involve the whole world, the affluent and the poverty-stricken nations alike. Abraham Lincoln's saying during the American Civil War that "This nation cannot long survive half slave and half free" now applies, in my judgment, to the world.

This world cannot long survive one third affluent and two thirds on the road to perdition. This is the challenge, and we do not need new research to be able to put the existing knowledge that science and technology has already provided to work to find a solution. All that is needed is the proper manpower, the proper capital, the proper organization, and the replacement of the present competition among the world powers by a new kind of co-operation against this common enemy. What science and technology have created in these past 50 years can give the world the power to save itself, and unless something is done soon, we will all go down. We should not seize population control as the answer. Population control is only one item in the dynamic cycle with which we have to deal.

It will come with the development of health programmes among the peoples, with education, and with development of other things. To concentrate on population control and to forget about economic and agricultural development, and the people needed to bring about this development, would be a grave mistake. We are going to need new armies to replace the present ones, armies of teachers, road builders, architects, managers, and all kinds of engineers. These will be the armies coming not as colonialists, not as conquering powers, but as a new crusade, a crusade that will set on fire the imagination of young people who are now lost in a world where they cannot find anything to dedicate themselves. This is the message that science and technology should carry to the world.

The humanist scientists among us should keep calling on the world to remember that unless something is done, that unless the present way of doing things is changed — we shall all go down the drain. This is the message we should carry, that the power is here, but that science and technology cannot act alone. We must work together, and things must be changed soon, before it is too late.

Appendix

The Weizmann Institute
By Professor Albert B. Sabin, President

The Weizmann Institute of Science in Rehovot, Israel, is a unique institute of scientific research and post-graduate education.

Located in a small country under siege, the Institute has become an important partner in the international scientific community by contributing to the sum total of human knowledge on which the future of mankind depends.

Today, the Institute is in the forefront of research in the life sciences, physics, chemistry and mathematics, and has become an important scientific resource not only of Israel but also of the world.

In the present era, it is our plan to supplement the excellent fundamental research, which has won the Institute its present eminence, by an expanded programme of mission-oriented research specifically designed to deal with important human problems, such as the conquest of cancer and other diseases, control of environmental pollution, desalination, development of new sources of energy, and new materials for industry. In all of these activities the Institute serves not only Israel but the entire world.

Excellent scientific research, of great importance not only for increasing our understanding of the world we live in, but also for providing the knowledge for the solution of some of the most critical problems facing mankind, is being performed in many universities in different parts of the world. In universities the primary task is to transmit the accumulated knowledge to new generations of students, while the discovery of new knowledge is a secondary task.

At the Weizmann Institute of Science, on the other hand, discovery of new knowledge is the primary task, and the training of new generations of scientists is the accompaniment of the search for knowledge. The Weizmann Institute is unique because research in all the natural sciences is being pursued on a single campus by highly competent scientists. The fact that they live together in a closely knit community results in a maximum cross-fertilization of ideas among the different scientific disciplines — a factor of the greatest importance in creating the innovations and new methodologies that are so essential to progress in all fields of scientific endeavour. At the Weizmann Institute, as at important universities, there is also the responsibility for "public service", i.e. to use the special competence of the faculty to work on problems of human importance to the community, the nation, the region, the world. Founded as the Daniel Sieff Research Institute in 1934, it became the Weizmann Institute in 1944, as a 70th birthday gift to the scientist-statesman, Dr. Chaim Weizmann, who subsequently became Israel's first President.

Leaders of the international scientific community have recently ranked the Weizmann Institute as "one of the world's truly eminent scientific research institutions . . . an achievement probably unparalleled in history." John F. Kennedy, one of the many renowned Honorary Fellows of the Institute, when he was President of the United States,

Appendix

said: "The Weizmann Institute has already, in a few years, earned its greatness. There are few institutions that have made such an extraordinary and enduring achievement in so short a time."

In 1971, the total staff at the Institute comprised about 1,900 people. This number included scientists, engineers, graduate students for the Ph.D. and M.Sc. degrees, technicians, service and administrative personnel. It also included about 100 visiting scientists who spent from six months to one year or more working in the 19 departments of the Institute. The Feinberg Graduate School, with 515 students, is chartered by the Board of Regents of the State of New York, and is recognized as an "American School Abroad" which has an extensive interaction with the foremost institutions of higher learning in the United States. American and European professors regularly come to our Institute, and members of the Weizmann Institute faculty are invited to serve as visiting professors abroad. The Institute interacts extensively with institutions of higher learning in the U.S.A., Great Britain, Europe and Latin America.

Some 400 research projects are now in progress at the Institute. Just to name the departments conveys the broad spectrum of scientific endeavour:

Life Sciences: Cell Biology; Experimental Biology; Biological Ultrastructure; Biodynamics; Genetics; Plant Genetics; Chemical Immunology; Biochemistry; Biophysics; Polymer Research.

Physics: Nuclear Physics (Low Energy, High Energy, Solid State); Electronics.

Chemistry: Organic Chemistry (Solid State, X-ray Crystallography, Photochemistry, Synthetic Chemistry, Materials Research); Isotope Research; Chemical Physics; Plastics Research.

Mathematics: Applied Mathematics (Computer Science, Geophysics, Oceanography, Systems Engineering); "Pure" Mathematics (including mathematical genetics, ecology and celestial mechanics).

The Department of Science Teaching, not listed above, is concerned with the development of up-to-date curricula, new textbooks, and science teaching aids that are designed to bridge the present long time gap between the rapid growth of scientific knowledge and its transmission to high school teachers and students. It is one of the manifestations of social responsibility so evident among the scientists at the Weizmann Institute.

Recently, four departments of the Institute were joined in a new multi-disciplinary cancer research centre which will give special emphasis to problems with a direct bearing on human cancer.

To assure that our research will be maintained without interruption, and that the promise it holds for contributing to human welfare may be fulfilled, we need more than ever increasing support from friends all over the world.

Name Index

Adams, Henry Carter 27
Adams, John C. 64
Amalrik, Andreï 175
Aristotle 113, 194, 209
Aron, Raymond 103, 110, 111, 167

Bach, John Sebastian 3
Bachelard, Gaston 117, 118, 176, 181
Bacon, Francis 167, 171
Bard, B. J. A. 94, 97
Beethoven, Ludwig van 199
Blum, Léon 185
Bohr, Niels 61, 83, 98, 99
Boyle, Robert 49, 50, 84, 170
Brecht, Bertold 58
Brunetière, R. Aron 90
Brunner, Jerome 59

Casimir, Hendrik 60, 61, 96, 104
Chardin, Teilhard de 23
Charney, Jule 75, 81, 82
Cherwell, Lord 79
Churchill, Winston S. 79
Cohn, Josef 3
Cramer, Friedrich 16, 18, 19, 30, 31, 33, 47, 50, 51, 55, 92, 94, 134, 147, 205, 206

Dart, Raymond 190
Darwin, Charles 20, 42
Delgado, José 146
Descartes, René 167
Dobson, G. M. B. 79

Duprée, A. Hunter 168, 181
Durkheim, Emile 39

Eban, Abba 162
Eckmann, Beno 93, 103
Edison, Thomas 81
Eiberfeld, Eibel 41
Eigen, Manfred 22, 46, 55, 58, 99
Einstein, Albert 114, 200

Feinberg, Abraham 142, 239
Feldman, Michael 182, 183, 210
Frankfurter, Felix 126
Freud, Sigmund 28, 40, 145, 189

Galbraith, John Kenneth 174
Galilei, Galileo 58, 59, 96, 175
Galle, John G. 64
Gentner, Wolfgang 129, 130, 131, 143
Gide, André 159
Gilgamesh 24
Gorki, Maxim 85
Gregory, David 84

Haydn, Franz Joseph 3
Hegel, Georg W. F. 19, 120
Herodotus 118
Hersch, Jeanne 54, 59
Higatsberger, Michael J. 89, 164, 165
Hilgard, Ernest 149
Hobbes, Thomas 121
Hornig, D. F. 76
Hozelitz, K. 191, 202

Name Index

Ilitch, Ivan 140

Jacob, François 135
Jaspers, Karl 181
Johnson, Lyndon B. 76

Kant, Immanuel 65, 120
Katzir-Katchalsky, Aharon 31, 32, 33, 46, 51, 53, 54, 57, 139, 144, 146, 152, 153, 154, 155, 157, 159, 162, 195, 207
Kendrew, John C. 13, 14, 15, 30, 49, 143, 209
Kennedy, John F. 4, 238
Ketelaar, J. A. A. 96, 102
Khorana, Hargobind 36
Kleist, Heinrich von 30

Leary, Timothy 40
Lefèvre, Théo 5, 6, 7, 11
Leverrier, Urbain 64
Liebig, Justus von 26
Lighthill, James 50, 87
Lincoln, Abraham 210
Lindemann, Frederick Alexander (Lord Cherwell) 79
London, Perry 148
Lwoff, André 181

Maaløe, Ole 131, 132, 133, 161, 208
McDonald, James 78, 82
MacGillavry, Carolina 54
Magruder, William M. 80
Malfatti, Franco-Maria 107
Malraux, André 126
Malthus, Thomas Robert 26
Mandelstam, Ossip 84, 85
Manuel, Frank E. 85
Marcuse, Herbert 28, 29

Marx, Karl 19, 179, 183, 184, 189
Maslow, W. 44
Medvedev, Zaures 175
Melzak, Ron 149
Milner, Alfred 146
Monod, Jacques 135

Newell, Homer E. 79
Newmann, John von 75, 76
Newton, Isaac 49, 50, 63, 64, 65, 66, 83, 84, 85, 167, 170
Nixon, Richard 100, 114

Oldenburg, Henry 49, 84
Olds, James 146
Oppenheimer, Robert J. 168, 172, 177
Orwell, George 194

Pauli, Wolfgang 61
Pekeris, Chaim L. 49, 50, 72, 73, 87, 89, 91, 98, 99, 157, 163, 208, 209
Perutz, Max 14
Petit, Raymond 190
Piaget, Jean 38, 43, 59
Picard, Leo 86
Picht, Georg 29
Polanyi, Michael 197
Press, Frank 76, 77
Price, Derek de S. 160
Prigogine, Ilya 43
Proxmire, William 80, 82

Rebeaud, Laurent 158
Rickover, Hyman George 81
Roosevelt, Franklin D. 114
Rousseau, Jean-Jacques 170
Ruszak, Theodor 39
Rutherford, Ernest 49

Name Index

Sabin, Albert B.   1, 2, 3, 50, 63, 95, 99, 104, 143, 195, 203, 237
Salomon, Jean-Jacques   166, 167, 190, 195, 210
Samuel, David   131, 142, 143, 152, 157, 159, 160, 162, 208
Sargeant, John   159
Scholem, G.   49, 152, 159, 160, 161, 162, 200, 207
Shalit, Amos de   131
Singer, Fred   79, 80, 81
Skinner, B. F.   147
Snow, C. P.   80
Spinelli, Altiero   10, 11, 99
Stalin, Josef   115

Tagore, Rabindranath   45
Theseus   24
Tribus, Myron   79

Van Cotthem, W.   53
Van Hove, Léon E. P.   62, 63, 102, 160, 207
Vavilov, Nikolai Ivanoch   175
Vian, Boris   178, 181
Vivaldi, Antonio   3

Wagner, Richard   120
Wald, George   80
Warburg, Siegmund G.   105, 106
Watson-Watt, Robert   14
Weisskopf, Victor F.   52, 59, 91, 94, 98, 99, 108, 131, 160, 161, 192, 193, 203, 290
Weizmann, Chaim   108, 109, 111, 118, 122, 126, 203, 238
Wells, H. G.   30
Wiener, Norbert   157, 158, 163
Wrangel, Peter Nicholaievich   84

Subject Index

Acceleration, Law of 27
Accelerators, atomic 69, 97
Activity, mental 150
Acupuncture 149
Advertising 148
Aggression 146
Agricultural settlement 86
Aircraft, jet 51
Air Force 115
Alchemy 49, 50, 84
America, see United States of America
American Physical Society 81
Amino acids 20, 21
Ammonia-CO_2 atmosphere 20
Amos de-Shalit Science Teaching Center 131
Amphibia 147
Analysis, numerical 76
Anaemia, pernicious 26
Anthropology 43, 166
Antibiotics 90, 137
Apes 25, 190
Apostles 180
Arab countries 120, 125, 127
Arms 114, 126, 181
Art 186
Asceticism 29, 30, 47, 48, 206
Astronomy 58, 64
Astrophysics 68, 71, 207
Atmosphere 20, 75, 92
  ozone 79
Australopithecus africanus 190

Atoms 26, 89, 169, 175
Atomic bombs 112, 113
  explosion 26, 74, 113, 114
  physics 35, 72, 175, 202
  weapons 28, 114
Automata theory 93

Bacteria 21, 26, 113, 133, 134, 135, 136, 137, 161, 208
Barbarism 204
Behaviour 27, 42, 142, 189
  ritualistic 42
Behaviour Control 148
Belgium 5-7, 11, 62
Betrayal 123
Biochemistry 26, 239
Biodynamics 239
Biological history 22
  norm 30
  problems 35
  research 16, 25, 26, 27, 88
Biologists 41
  dreams 50
Biology 32, 41, 112, 143, 157, 182
  experimental 239
Biophysics 35, 239
Biosphere 22, 94
Blood groups 26
Boolean algebra 93
Boyle's Law 49
Brain 35, 134, 144, 146, 147, 149, 150, 151, 152, 157, 199
  alpha rhythms 148

Subject Index

cells 159-160
chemistry 142, 144
computer interaction 151
control of 37
damage 152
electrical stimulation 146, 151
functions of 145
immortal 25
power 24
research 38, 56, 143, 151, 152, 157, 159
transplant 147
washing 145, 159
British Council 182
Bulletin of Atomic Scientists 113

Cancer 2, 78, 79, 82, 100, 101, 237, 240
Capital 174, 187
Capitalist economy 115, 183
"Care poles" 25
Cartesian formula 112
Cat 146
Catalysis 21, 135, 190
Category theory 93
Causality 19, 57, 64, 67, 71
Causa vitae 23
Cavendish Laboratory, Cambridge 14, 49
Cells 20, 21, 47, 239
 division 21, 134
 nerve 147, 150
Central nervous system 150
CERN 62, 63, 69, 91, 96, 98, 102, 130, 131, 192
Challenge of the future 4, 16, 19, 31, 37, 38, 56, 70, 122, 139, 143, 170, 205, 206, 210
Chemical Immunity 26, 163, 239

Chemical Physics 240
Chemistry, organic 240
China 28, 149, 198
Cholesterol 90
Chromosomes 35, 36
Church, catholic 44
Citizenship, global 38
City 179, 196
Civilisation 29, 119, 206
Class struggle 183, 184
Code, ethical 152
Cognition 39, 43
Commission of European Communities 10, 11, 107
Communications 8, 41, 91, 119, 138
Compassion 203
Computors 67, 75, 93, 133, 150, 157, 191, 208
Concorde 78, 88
Condon report 82
Conflicts 179
Congress, U.S.A. 79, 100
Contraception 26, 178
Contrat social 28
Control loops 135, 136
 mechanisms 138, 139
Cortisone derivatives 90
Cosmos 154
Countries
 developed 34, 101
 developing 42, 52, 53, 57, 91, 93, 95, 169, 210
Creation story 19, 20
Creativity 151
Crime 35, 84, 144
Crusade 211
Curiosity 203, 205

Dangers 151, 157, 161
Data-processing 76
Death 55
Decision-making 150, 187, 188, 189
Defence, national 115, 119, 120, 126
Dehumanization 200
Democracy 36, 183, 185, 186, 187
Deoxyribonucleic acids (DNA) 20, 138
Desalination 237
Desert 86
Destiny 29
Deterioration 169
Determinism 66, 67
Developments, non-material 56
Diabetes 26
Dialectical relations 155
Diaspora 122, 123, 125, 126
Differences, ideological 184, 185
Dilemma 47
Diseases, hereditary 36
 infectious 26
"Dissipative structures ..." 43
Double Helix 20
Dreams 50, 151
Drugs 137, 144, 145, 149, 152
 addictive 144
 culture 144
 mind-affecting 145
Dynamics, fluid 76

Earth 169
Earthquake 74–77
 Alaskan 76–78
 Prediction 77–78
 San Fernando 77
Economics 7, 8, 9, 26, 28, 33, 89, 93, 95
Economy 7, 93, 97

Education 8, 44, 57, 93, 96, 131, 140, 143, 148, 150, 200, 201, 210, 237
 of Scientists 53
 of Society 53
Efficiency 90, 134
Efficiency-principle 162
Egypt 97
 Ancient 19
Electric chair 81
 power transmission 81
 shock 148
 stimulation 146
Electricity generating stations 112
Electro-analgesia 149
Electrodes, implanted 146
Electrodynamics 201
Electronics 67, 239
Electron microscope 207
EMBO 131
Energy 27, 154, 161
Engineering 186, 211
Entropy 66
Environmental problems 9, 19, 22, 25, 37, 81, 87, 88, 95, 169, 195, 196, 197
Enzymes 18, 134, 135, 136, 149
Epi-genetics 139
Eros 28, 29
ESB 146, 147
Ethical behaviour 43
 conclusions 41
 dictum 36
 philosophy 39
 problem 37, 157
 values 184, 188, 189
Ethics 152, 179
Ethnology 123
Europe 11, 12, 15, 26, 92, 118, 167, 200, 201, 239

Subject Index

European Economic Community 12, 87
Evolution 22, 23, 24, 25, 28, 42, 44, 46, 47, 51, 55, 58, 134, 135, 136, 137, 138, 139, 160, 161, 209, 225
Evolutionary concepts 57
   process 30, 208
   teleonomy 22
Exploration, scientific 4
Exponential acceleration 27
   growth 196, 206

Farmers, sedentary 24
Faustian Man, competitive principle of 28
Feinberg Graduate School 142, 182, 239
Fertilizers 162
Fish, parthenogenic 147
Forecasting, numerical 76
France 88, 120, 123, 124, 185
Freedom 29
Freethinkers 127
Free will 147
French Committee of the Weizmann Institute 90
Fusidic acid 137, 138

Garden of Eden 204
GARP 76
Gedankenexperimente 58
Genes 36, 51, 149, 157, 189
   controlling 178
Genesis, book of 39, 204
Genetics 20, 21, 36, 42, 48, 151, 157
   plant 239
Geologist 86, 197
Geophysics 72, 77, 78
Germany 53, 209, 131
Gestalt 58

Globe 51, 58, 87, 154, 161
God 19, 58, 189
Gold, with Quicksilver, Incalescence of 49, 84
Golems 157
Governments 7, 11, 15, 175
Great Britain 23, 88, 184, 239
Greeks, science of the 59, 155, 167
   history 19
   philosophy 109
Green Revolution 161
Gross national product 100
Groundwater wells 86
Growth, minimum level 95

Hallucinogens 145
Happiness 168
Heart 149
   rates 148
   transplantation 147
Heredity 24, 169
Hippies 40
Hippocratic Oath 152
Hiroshima 67, 112, 168
History 138, 139, 170, 173
Holy Books 122
Homo sapiens 19, 20, 24, 51
Hormones 25
   controls 138
   sexual 26
Humanists 8, 205, 208
Humanity 8, 122, 143, 203
Hunger 24, 161, 162
Hunters, nomadic 24
Hygiene 8, 24, 113
Hypertrophy of curiosity 39
Hypnosis 149
Hypocrisy 119

Immunology 147, 182
Impact of science on Israel 33
Incalescence 49, 50, 84
Income, per capita 27
Industrial consulting 83
    nations 102
    research 95
    revolution 122, 184
    system 174, 177
Intellectualism, anti- 39
International relations 7, 37, 38
Isotope Research 240
Israel 7, 15, 32, 33, 86, 92, 108, 109, 120, 122, 123, 124, 126, 131, 155, 162

Japan 77, 198
Jet engines 88
Jewish family 123
Jewish State 122, 126
Judaism 34, 123

Kinetics of progress 27
Knowledge 112, 167
    Moratorium on 170
    tree of 39, 56

Laissez-faire 48, 177, 180
Language 125, 139, 147
    Universal 34
    Western 34
Latin America 140, 239
Leadership 76
Life, origin of 20
Life sciences 175
LSD 40

Machines 64
Malnutrition 161

Manhattan Project 192
Mankind 4, 28, 29, 37, 39, 43, 44, 45, 58, 59, 84, 119, 169, 196, 203, 237, 238
Market economy 177
    limited 102
Marxism 23
Mass conditioning 148
Mass media 148
Mathematics 60, 93, 103, 176, 179, 237, 240
Maturity 40, 44
"Meaning-of-meaning" 54, 55, 59
Mechanics, classical 64, 65, 66
    Newtonian 63, 64, 65
    of reactions 142
    relativistic 65
Medicine 8, 26, 137, 194, 209
    achievements 30
    experimental 16
Memory 150
Metamorphosis 27
Metaphysics 176
Methane 80
"Military-industrial complex" 174
Misera plebs contribuens 155, 156
Model 138, 208
Molecular biology 20, 30, 35, 169, 173
    genetics 38
    level 46
    reproducing mechanisms 22
Money 100, 102
Monkey 146, 148
Moon 169
Morality 38, 39, 40, 55, 71, 158
    of Science 45
Moratorium of Science 49, 50, 52, 53, 55, 56, 58, 89, 101, 131, 143, 144, 151, 170, 195, 209

Subject Index

Mortality, infant 16, 26
Music 186
Muslims 127
Myoglobin 14
Mystics 78

Nagasaki 112
Naïveté 119
Natural history 24
    Selection 20, 21
Nature 40, 59, 194, 197, 198, 200
Nazis 53, 209
Near East 19, 119
Neptune, planet 64
New Atlantis 171
"New Mathematics" 93
Newspapers 91, 111, 124
Nihilism 168
Nobel Prize 14, 180
Noise 87, 112
"Non-proliferation pact" 28
NORSAR 74
Nuclear fission 67
    physics 49, 67, 112, 151, 202, 239
Nucleic acid chemistry 18, 20, 21
Nucleus 21, 36
Nutrition 25, 26

OECD 164, 166, 167
Ontology 19, 22
Operant-conditioning 147
Oppenheimer case 177
Organisms, living 20
Original sin 30
Oxygen 20, 25, 149, 150
Ozone 79, 88

Pan-Germanism 120
Parenthood 25

Penicillin 56, 90
Personality 25, 152
    problem 43
Philips 191
Philosophers 40, 100, 158, 204
    Stone 50, 84
Physical control of Mind 146
    Mind 44, 143, 145, 150, 152, 157
Physics 160
    high energy 68, 69
    low temperature 60
    particle 202
    reactor 164
    semi-conductor 67, 143, 262
    theoretical 60
    thermo- 201
Physio-mathematical theory 22
Pioneering scientists 45
Planet 24, 64, 294
Planning 11, 28, 29, 30, 99, 100, 104, 173, 206, 207, 238
    future 161
    local and national 161
Poetry 84, 85, 158, 159, 163, 186
Politics 48, 54, 68, 71, 113, 114, 115, 116, 119, 158, 168, 170, 171, 172, 173, 176, 177, 179, 180, 185, 186
Pollution 28, 38, 92, 95, 97, 198, 237
    crisis 167
    stratosphere 79
Polymer Research 239
Population 8, 28, 46, 87, 196
    control 210
    explosion 16, 26, 52, 196
    growth 26, 27
    Israel 33
    Over- 52, 178
    World 27, 194

Subject Index

Positivism 179
Power, Military 117, 174
  escalation of 169
  purchasing 184
Pre-hominids 19
Primeval soup 30
Priorities 92, 99, 101, 102, 108, 131, 143, 153, 202
Production 184, 206
Progress 27, 28, 29, 30, 42, 92, 107, 168
Proletariat 183
Propaganda 113, 148
Prophets 180
Protein 14, 20, 136
  control 136
  primitive 21
  synthesis 21, 137
"Prussians of the Middle East" 125
Psyche 27, 59
Psychoanalyst 163, 203
"Psychobiology" 143, 149
Psychology 38, 43, 148, 158, 163, 196
Psychopharmacology 144, 151
Psycho-sociology 176
Public, the 199
  investments 177
  opinion 7
Pugwash Conference 113

Radar 14
Radioactivity 79, 150, 202
Realitätsprinzip 40, 54
Reason 118, 169, 180
Religion 123, 180
Renaissance 8
Repression mechanisms 135
Reproduction 22, 23
Republic of Israel, president of 111, 122

Research 8, 9, 11, 12, 91, 103, 104, 115, 118, 143, 144, 149, 151, 176, 178
  autonomy of 177
  basic 56, 197
  biological 204
  budget of 117
  fundamental 209, 237
  mission-oriented 237
  pure 15, 67, 154, 155
Resources 97, 98
Responsibility, moral 42, 163
Retardation, mental 149
Ribosomes 134, 135, 208
Rice, high yield strains of 161
RNA 21
Robot 32, 146, 147, 157
Royal Society 49, 50, 81, 83, 84
Russian TU-133 78

Satellites 76
Salt 66
Scarcity 28, 29, 52
Scepticism 168
Science
  anti- 144
  arrogance of 199
  basic 53, 194, 197, 198, 201, 202
  Big 68, 70, 102, 116–117
  cathedral of 97, 103, 169, 180
  conscience of 181
  dominated society 35
  fiction 37, 147
  fundamental 108, 160
  government 168
  history of 171
  ideology of 170, 180
  Israel 34
  materials of 175

Subject Index

methodology 142
misuse of 180
neutrality of 34, 35
policy 71, 94, 103, 166, 173, 207
politics 112
teachers 201
teaching arts 240
technicians 156
war 52
Science and Public Affairs 113
Science et Politique 166
Science in the Federal Government 181
Scientification 43
Scientific Youth of Belgium 190
Scientist's authority 40
 community of 121
 curiosity of 39
 madness 146
Secrecy 70, 121
 military 120
Seduction, romantic 148
Senate, Belgium 6
 U.S.A. 78, 79, 81, 85
Set theory 93
Sexual deviants 148
 repressions 189
Sharm-esh-Sheik 121
Shell 191
Simonists 180
Social problems 33
 progress 9
 sciences 67, 122
 system 161, 162
Society, Impact of Science on 3, 7, 9, 11, 38, 51, 63, 78, 81, 83, 86, 90, 92, 95, 107, 108, 133, 155, 161, 167, 181
Society, Science-dominated 35
Sociologists 43

Sound, speed of 88
Soviet Union 85, 115
Space 175
 age 89
 time and 65, 68
Spinal cord 149
Standard of living 57, 101
Statistics 66, 71, 156
Steroids 90
Stimulants 144
Stone Age, Early 24, 51
Stratosphere 79, 80
Students 119
Suez 123, 124
Sun 96
Super-conductivity 68
Super-sexual satisfaction 37
Supersonic Transport 78, 82, 87, 88, 91, 92, 170, 202
Swiss Federal Institute of Technology 93, 192
Synchrotrons 69
Systems, analysis 162
 closed 38, 139, 140, 154, 161
 open 139, 140
 symbolic 160

Teaching 140, 156, 211, 240
Technical colleges 191
 progress 7
Technological decisions 88
 developments 101, 195
 innovation 9
 Future, Challenged 16
 planning 30
 progress 184, 186
Technology 29, 31, 33, 52, 64, 71, 97, 100, 113, 194, 195, 180

Technonature 174, 175
Teleonomy 22, 23, 24
Telephone 150, 157
Telescope 58, 96
Television 25
Temperatures, low 68
Territorial perception 38
Test Ban Treaty, Partial 74
Thermodynamics 23, 32
Third world 140, 141
Tracer methods 202
Transfer ribonucleic acids (RNA) 21
Transistor 67, 112
Trans-national community and city 119, 121
Transplants, organ 25, 26, 48, 152
Transportation 51, 80, 92, 133, 162, 196
Truth 59, 112, 119, 173, 175, 176, 178, 180, 188, 189, 209

U.F.O.'s 82
United Nations 57
United States of America 3, 4, 9, 44, 69, 78, 92, 100, 101, 115, 118, 140, 167, 168, 185, 195, 196, 200, 201, 202, 209, 239
  Civil War 210
  Congress 91
  Department of Defence 115
  Department of Transportation 79
  President of 114, 238
  Science Advisor 82
  Supreme Court of 126
Universe 4, 23, 96, 114, 154, 198, 200
Universities 115, 118, 191, 197, 238
  students in 44

University of
  Amsterdam 96
  Basle 110
  Berlin 192
  Brussels 31, 62, 110
  California 32, 47, 142, 182
    Radiation Laboratory of 130
  Cambridge, England 18, 49, 90, 188
  Cincinatti 2
  Columbia 110
  Copenhagen 131-2
  Cornell 110
  Darmstadt 18
  Edinburgh 182
  Erlangen 130
  Frankfurt/M 130
  Freiburg 130
  Geneva 54
  Ghent 6, 53
  Graz 164
  Harvard 110
  Heidelberg 18, 130
  Jerusalem 32, 86, 87, 142
  Johns Hopkins 10
  Leyden 60
  Maryland 79
  Miami 79
  Minnesota 164
  New York 2
  Oxford 188
  Paris 110
  Princeton 62
  Rochester 192
  Rome 10
  Sorbonne 166
  Southampton 110
  Southern California 148
  Stanford 182

Tübingen 110
Utrecht 62
Vienna 164
Yale 146
Uranus, orbit of 64
Urban centres 8
 politics 179
Utopian Golden Age 28

Vaccine, oral polio 2
Value judgments 179
Vietnam 115, 116, 138, 167
Viruses 2
Vitalism 30
Vitamins 145

War 126
 Apocalyptic 112
 Second World 52, 56, 126, 159, 168, 172, 173, 209, 238
 Six Day 125, 126
Water 8, 66
 vapour 79, 80
Weapons 67, 194, 209
Weather, Climate Modification 82, 85
 forecasting 75
Weizmann Institute of Science 1, 2, 3, 15, 31, 32, 69, 72, 86, 106, 107, 108, 109, 112, 122, 127, 131, 147, 182, 203, 237, 238, 239, 240
Wheat, high yield strains of 161
Wings, variable sweep back 88
Writers 84

Year 2000 78
Youth 39, 53, 131, 150, 168, 190, 197

Zeitgeist 23